Membrane Based Point-of-Use Drinking Water Treatment Systems

Need for point-of-use water treatment systems (Painting by Anjali Kothe, Project Associate, CSIR-NEERI, Nagpur, India)

Membrane Based Point-of-Use Drinking Water Treatment Systems

Pawan Kumar Labhasetwar and
Anshul Yadav

Published by   **IWA Publishing**
Unit 104–105, Export Building
1 Clove Crescent
London E14 2BA, UK
Telephone: +44 (0)20 7654 5500
Fax: +44 (0)20 7654 5555
Email: publications@iwap.co.uk
Web: www.iwapublishing.com

First published 2022
© 2022 IWA Publishing

Disclaimer

British Library Cataloguing in Publication Data
A CIP catalogue record for this book is available from the British Library

ISBN: 9781789062717 (Paperback)
ISBN: 9781789062724 (eBook)
ISBN: 9781789062731 (ePub)

This eBook was made Open Access in January 2023.

© 2023 The Authors.

Contents

Chapter 2
Water supply systems and the need for point-of-use treatment systems . . . 29

Chapter 5
Modelling membrane operations in membrane-based point-of-use
water treatment systems ***105***

Chapter 6
Operation and maintenance of membrane-based point-of-use
water treatment systems ***131***

Chapter 7

Techno-economic analysis of membrane-based point-of-use water treatment systems

Chapter 8
Certification and evaluation of membrane-based point-of-use water treatment systems **167**

About the authors

Dr Pawan Labhasetwar, Chief Scientist, Water Technology and Management Division, National Environmental Engineering Research Institute (CSIR-NEERI), has obtained PhD in Environmental Engineering. He has excellent scientific contributions in the area of Environmental Engineering and Water Treatment and Management. Dr Labhasetwar is the First Chairman of the Committee of Bureau of Indian Standards (BIS) for developing standards for "Reverse Osmosis (RO) Based Point-of-Use (PoU) Water Treatment System - Specification" and familiar with membrane processes. Dr Labhasetwar is also Head of the World Health Organisation's Collaborating Centre for Water and Sanitation.

Dr Anshul Yadav is Scientist in Membrane Science and Separation Technology division, CSIR-Central Salt and Marine Chemicals Research Institute, Bhavnagar, Gujarat. Dr Yadav received B.Tech.-M.Tech. dual degree in Mechanical Engineering from Indian Institute of Technology Kanpur and Ph.D. in Engineering Sciences from Academy of Scientific and Innovative Research, New Delhi. Dr Yadav has excellent scientific contributions in the area of water treatment specializing in membrane-based (waste)water treatment techniques. Dr Yadav's research interest also include computational modelling and simulation of membrane and adsorption based water treatment processes, gas sensing, and gas separation.

Foreword

The availability of safe drinking water is a human right and a public health priority. Nevertheless, we know from global monitoring data that in 2020 nearly 800 million did not have access to a basic water supply, and two billion did not have access to 'safely managed water. The Sustainable Development Goals (SDG) target 6.1 to achieve universal and equitable access to safe and affordable drinking water by 2030 is unlikely to be reached. This will mean continued high levels of preventable disease from pathogens and chemicals found in contaminated drinking water. Given the limited progress in providing access to water supplies, whether managed by utilities or communities, there has been sustained interest in point-of-use water treatment as a way of rapidly increasing access to safe water for households across the globe. While such technologies will not prevent all WASH-related diseases and the importance of improving sanitation and hygiene remains clear, water treatment does help reduce exposure to pathogens and harmful chemicals and contributes to better health.

This book is focused on the use of membrane filtration in point-of-use water treatment, recognising the importance of such technologies in providing accessible, effective water treatment. The book is timely and is an accessible introduction to processes often considered the preserve of specialised engineers. It provides a comprehensive overview of the technologies, their management, the economics, and the regulation of membrane filtration units. The authors provide students and practitioners with a comprehensive and thorough analysis of membrane filtration units, providing insights into an important global technology.

The authors have written a well-structured book that allows readers to develop their knowledge and understanding of membrane filtration. The first chapter starts by setting out the types of water sources used for drinking supplies and the range of water quality parameters that must be addressed. Importantly this includes a range of emerging contaminants that will become increasingly important in the coming years. We then move into chapter two, which describes water supplies and discusses conventional water treatment processes and their use to tackle specific threats. This sets the scene well for chapter 3 where we now learn about the different types of point-of-use systems and their effectiveness. This chapter allows the reader to understand how membrane filtration systems work and how they compare to other forms of point-of-use systems.

Chapter 4 describes the design of membrane filtration point of use systems, which is then complemented in chapter 5 by a comprehensive overview of the modelling of membrane systems and chapter 6, covers the operation and maintenance of membrane systems. These three chapters are the heart of the book and give the reader a good understanding of how membrane systems work, the key characteristics of design and construction, how to model their functioning, and how they must be operated and maintained. These are all important in the understanding of technology.

This understanding is critical as in chapter 7 we are taken through the key technical and economic considerations around the deployment of membrane point-of-use systems. This provides information about cost-effectiveness, drivers of costs (including capital investments), and information about the global market for membrane point-of-use systems. The final chapter deals with the regulation and, in particular, certification schemes for point-of-use systems. This has too often been overlooked by the importance of understanding how good certification systems work cannot be overstated.

In conclusion, this book is a very welcome addition to the water treatment literature. It addresses an important technology from technological, economic, and governance perspectives providing the reader with a good understanding of such systems can be deployed and regulated.

Professor Guy Howard,
University of Bristol, UK

doi: 10.2166/9781789062724_xv

Preface

Water and sanitation investments can result in a net economic benefit since the reductions in adverse health consequences and healthcare expenditures surpass the price of implementing the interventions. This holds true for significant water supply infrastructure investments and domestic water treatment. However, the ever-widening gap between global water demand and supply has compromised water sustainability and quality. Water, sanitation, and hygiene continue to be a problem for billions of people worldwide. According to the World Health Organisation, in 2019, 2.2 billion people lacked safely managed drinking water services (drinking water accessible on-premises, available when needed, and free of pollution), and 4.2 billion lacked securely managed sanitation facilities. Moreover, only 6 of 99 nations are on track to achieve universal (>99%) safe drinking water management by 2030, as reported by the joint monitoring committee, WHO 2021 report.

Subsequently, rapid urbanisation, industrialisation incessant pollution, and climate change have adversely degraded drinking water sources. Water resources are fast depleting and deteriorating. Water quality degradation at household/point-of-use (PoU) is mostly a result of diffuse-source contaminants and the spatial and temporal variability associated with these sources. Eighty per cent of the world's wastewater is dumped – largely untreated – back into the environment, polluting rivers, lakes, and oceans. Meanwhile, our drinkable water sources are finite. Water treatment methods are complicated when source water quality deteriorates, resulting in higher treatment costs and lower tap water quality.

Moreover, conventional water treatment plants are designed primarily considering the removal of suspended solids and microorganisms in most cases. In addition, water quality further degrades in transportation and distribution systems, particularly in the intermittent water supply. Several conventional water treatment processes are available such as coagulation, flocculation, precipitation, adsorption, ion exchange, electrochemical treatment, etc. Furthermore, these conventional water treatment plants suffer from several serious limitations such as complex operability, high maintenance requirements, energy inefficiency, high capital, operation and maintenance costs, and skilled personnel for operation. Difficulty in implementing water sanitation safety plans, particularly in developing countries, also restricts the delivery of safe water to the household. To alleviate growing concerns of unsafe water delivered through the non-protected catchments, increasingly contaminated water sources, insufficient conventional water treatment plants and highly porous distribution systems, particularly in developing countries, the PoU water treatment system is emerging as a preferred solution.

PoU water treatment systems to treat contaminated water primarily for drinking and cooking accommodate a small number of people, appropriate for short-term response and are low cost. The emergence of the PoU water treatment systems is primarily to provide water of reliable quality and treat a small quantity of water to meet the demand only for drinking and cooking purposes.

Membrane-based PoU water treatment systems such as ultrafiltration and reverse osmosis have proved to be a milestone in PoU water treatment eliminating various shortcomings of other water treatment technologies. Despite some of their limitations, membranes have emerged as a convenient and acclaimed treatment technology at PoU/households due to their versatility in removing almost all the contaminants from water. The main driver of a membrane-based PoU water treatment system is that it works without the addition of chemicals, with relatively low energy usage, and easy and well-arranged process conductions. Membrane surface modification has developed as a new technique to improve membrane performance in terms of better permeate flux (treated water) and lower fouling rate due to a weaker contact between fouling material and modified membrane surfaces for PoU applications. Plasma treatment, physical coating of a hydrophilic layer on the membrane surface, nanoparticles for surface modification, and chemical reactions on membrane surfaces are examples of such modification procedures. Hence with improving technology, there is an inevitable need to understand the basic operational parameters, design and maintenance of membrane-based PoU water treatment systems to eliminate the repercussions of lack of knowledge. There is also a need for techno-economic analysis of PoU water treatment systems to assess their economic viability, especially in developing countries.

The book describes membrane-based PoU water treatment systems and is divided into eight chapters. Each chapter covers issues related to water quality, water contamination, reasons for recent water quality degradation, conventional methods for water treatment – their limitations and the need for (membrane-based) PoU water treatment systems. Chapter 1 explains the sources and contaminants in drinking water (physical, chemical and microbiological). It briefly describes the effects of these contaminants on human health, aesthetics, etc., and the importance of removing these contaminants. Chapter 2 describes different units, advantages and limitations of the conventional water treatment plants and various PoU water treatment technologies. Chapter 3 covers components and membrane-based PoU water treatment systems and details about ultrafiltration, microfiltration, nanofiltration, and reverse osmosis-based systems. Chapter 4 describes the effective designing of various components of the membrane-based water treatment systems to make them more economical and practical. The design of pre-treatment and post-treatment and multi-stage/multi-barrier systems is also described in this chapter. Modelling and simulation, process optimisation, and case studies related to membrane-based PoU water treatment systems are included in Chapter 5. Chapter 6 includes operation and maintenance aspects, including pre- and post-treatment units. Techno-economic aspects of membrane-based PoU water treatment systems are elucidated in Chapter 7. Chapter 8 elaborates on national and international protocols for certification and system evaluation. The process of developing a national evaluation/certification protocol for PoU water treatment systems is also included here.

This book aims to cater to students/research scholars/academicians/ industries/practitioners in water science/engineering and environmental sciences. It can also serve as a reference for stakeholders concerned with water treatment. Considering classified/patented information and limited data about various components of membrane-based PoU water treatment systems, this book explores developing system design criteria based on discussions with manufacturers and experts. It can also be important for manufacturers and vendors of the systems as they can familiarise themselves with the scientific and technical details of the systems as important as commercial aspects. Government and non-government agencies and independent organisations interested in initiating certification and evaluation processes for these systems can also use this book. Given a broad set of possible readers who may find this book useful, an attempt is made to keep the scientific and technical explanation simple.

We hope that the readers will find this book useful.

Pawan Kumar Labhasetwar and Anshul Yadav

Acknowledgements

Writing a book on a very well-developed topic, which does not have much background literature, warrants many contributions. We have been fortunate to reach out to them and receive the necessary inputs. This book is about something we see and use every day, i.e., membrane-based Point of Use water treatment systems, yet we do not realize that not much is organized in designing these systems. We embarked upon the journey of completing this book, knowing fully well that several inputs would be required. Now, we completed the book, and these inputs are well received.

First and foremost, we wish to thank the Council of Scientific and Industrial Research (CSIR), Ministry of Science and Technology, Government of India, which supported us in writing this book through our respective institutes, namely CSIR-National Environmental Engineering Research Institute (CSIR-NEERI) and CSIR-Central Salt and Marine Chemicals Research Institute (CSIR-CSMCRI). We gained knowledge and experience in these institutes, which helped us put our thoughts through in the form of the book. Dr Atul Vaidya, Director, CSIR-NEERI, Nagpur and Dr Kannan Srinivasan, Director, CSIR-CSMCRI, Bhavnagar, provided the necessary support and requisite approvals and also ensured that we get sufficient time out of our busy schedule to write this book. In turn, we committed ourselves to complete the institute's assigned tasks and spending time writing the book.

Colleagues and friends are the pillars of strength, and their motivation always helps complete the desired tasks with responsibility. Dr Pranav Nagarnaik, Ms Neha Wachasunder, Ms Anjali Kothe and Ms Anupama Rodge from CSIR-NEERI, Nagpur; Mr Raj Vardhan Patel, and Ms Shweta Chaubey from CSIR-CSMCRI, Bhavnagar; Ms Pratibha Yadav and Ms Pushpmala Kuwer from IEHE, Bhopal and Mr Chandra Prakash Singh from IIT Kanpur provided direct inputs in the form of literature search and design details of membrane-based water treatment systems. Dr Sanjoy Roy, Orange City Water Private Limited, Dr Shweta Banerjee, Nagpur Municipal Corporation and their colleagues of respective organizations provided important photographs of the conventional water treatment plant at Nagpur, India. Dr Rajesh Lohiya, Incharge, Key

Resource Center, CSIR-NEERI, Dr Shibaji Ghosh, CSIR-CSMCRI, Bhavnagar and Dr Nikhilesh Trivedi, CSIR-CSMCRI, Bhavnagar were also supportive in processing our request of writing the book. The CSIR-CSMCRI PRIS number for this manuscript is 117/2022.

A book on such a topic cannot be completed without the help and inputs from many people. At the outset, we thank Mr Mark Hammond, Mr Niall Cunniffe, and other International Water Association (IWA) staff first for accepting our proposal to write the book. We would also like to especially thank Mark for his guidance over months during this long and enduring process. We are also grateful for helping us in editing and proofing the pages. We are equally grateful to the several people who provided us with the photographs for inclusion in the book. Their time and efforts were responsible for assembling these photographs. Dr Tapas Chakma, ICMR-National Institute of Research for Tribal Health (ICMR-NIRTH), provided photographs of fluorosis affected children. Dr Subhamoy Bhowmick, CSIR-NEERI, Kolkata and Prof Pankaj Roy, Jadhavpur University, Kolkata provided photographs of asreconosis affected people (we hid the faces of these people to avoid identification). Mr Ajay Popat, Ion Exchange India Limited, provided membrane-based PoU water treatment systems specifications. Dr Nitin Labhasetwar, CSIR-NEERI, Nagpur reviewed a few Chapters, and his inputs helped us believe that we have done justice with various Chapters.

We thank our colleagues Dr Girish Pophali, Dr Paras Pujari, Dr Gajanan Khadse, Dr Raghuvabnshi Ram, Dr C Padmakar, Dr Leena Deshpande, Dr Atul Maldhure, Dr Shalini Dhyani, Mrs Vandana Cinthray, Dr Parikshit Verma and Dr Lakshmikanthan from CSIR-NEERI, Nagpur and Dr Vinod K. Shahi, Dr Puyam S. Singh, Dr Amit Bhattacharya, Dr Vaibhav Kulshreshta, Dr Suresh K. Jewrajka, Mr Sanjay D. Patil, Dr Hiren D Raval, Dr Uma Chatterjee, Dr Rajaram K. Nagarale, Dr Nirmal K. Saha, Dr Saroj Sharma, Dr Santanu Karan, Dr Ketan R Patel, Mr Bhaumik Sutariya, Mr Pankaj D. Indurkar, Dr Mrinmoy Mondal, Mr Bipin Vyas from CSIR-CSMCRI, Bhavnagar for assisting directly and indirectly in compiling this book.

Dr Guy Howard, University of Bristol, kindly consented to write Foreword, which added value to the book. We also thank anonymous reviewers who kindly reviewed our book proposals, which improved the book's contents.

Further, we also thank our family members Dr Suvarna Labhasetwar (wife of Pawan Labhasetwar), Dr Prabhakar and Ms Sunita Labhasetwar (parents of Pawan Labhasetwar), Mrs Swati Gupta (wife of Anshul Yadav), Ms Manju Yadav (sister of Anshul Yadav), Mr Ram Karan and Anupama Yadav (parents of Anshul Yadav) for keeping our morale high during the writing of this book.

Disclaimer

The book is written and compiled by using the generic data available for the subject matter. Moreover, references are included wherever any data are used in the book. Design principles and criteria are derived based on the experience of authors, generic data available for the purpose and discussion with manufacturers and experts of point-of-use (PoU) water treatment systems. The authors attempted their best to analyse the data available globally and not proprietary in explaining the concepts and design of PoU water treatment systems. While effort is made to ensure the accuracy and completeness of information and data in compiling the book, human error or omission may be expected. Although the attempts are made to include the latest available data, some of the data used might be dated. There might be a resemblance with the available literature, which is purely coincidental as the authors took great care in the compilation of the book. Although professional experience and references were used in designing the systems, the authors make no assurance related to the design's success if the principles in the book are used due to scientific complexities, national/regional variability and ever-evolving characteristics of membranes and other components of PoU water treatment systems.

doi: 10.2166/9781789062724_0001

Chapter 1

Water sources and quality parameters

1.1 INTRODUCTION

Water is a renewable natural resource found in various forms throughout the ecosystem. It is vital for irrigation, industries, and domestic purposes and therefore indispensable for living beings and inanimate factors. Freshwater is crucial for the synthesis and structural formation of cell constituents and for transporting nutrients to cell and body metabolism. During civilisation, humans settled in places where plenty of water was available. However, as the population continued to increase, humankind started exploiting this natural resource for its benefit without even thinking about future consequences. There are countless reasons for water contamination, among which anthropogenic activities are of the greatest concern. Because of the callous behaviour of humans, water is getting more contaminated day by day, and the situation is becoming grim.

An adequate supply of safe and clean drinking water is a basic requirement for healthy individuals. Access to safe drinking water is a fundamental necessity. The most crucial factor in human development is access to clean water and basic sanitation services. Contrary to this, according to World Health Organisation (WHO) and United Nations International Children's Emergency Fund (UNICEF) 2020, nearly half of the world's population lacks access to adequate drinking water and sanitation. One out of every three individuals around the globe does not have access to potable water (WHO, 2020). There are two major issues in having clean and safe drinking water access. The first is the reduction in water availability due to increasing population and competitive demand, and the second is the anthropogenic activities leading to contamination of water sources.

The United Nations (UN) adopted sustainable development goals (SDGs) in 2015 to expedite actions to eradicate poverty, protect the planet, and ensure that all people live in peace and prosperity by 2030. The integrated 17 SDGs attempt to ensure environmental, social and economic environmental sustainability.

The SDGs were developed after the intense, comprehensive and inclusive dialogue in UN history and adopted by 193 countries. Although challenging, acceptance of SDGs by many countries have motivated people from across the globe that we can live in harmony with nature and with dignity. Achieving the SDGs by 2030 will be possible with ingenious efforts and partnership. SDG 6, related to 'Clean Water and Sanitation', is an important goal and is meant to provide access to safe water and sanitation to everyone across the globe. Although significant progress is made in improving access to safe drinking water and sanitation, many (billions) of people, particularly in rural settings, can still not avail of these essential services.

The following facts and figures reported by the UN highlight efforts needed to ensure access to safe water and sanitation (www.un.org/sustainabledevelopment/water-and-sanitation):

- Between 1990 and 2015, the proportion of the world's population who had access to better drinking water increased from 76% to 90%.
- Three out of ten individuals do not have access to safe drinking water, and six out of ten do not have access to safe sanitation facilities.
- One in four healthcare facilities lacks basic water services.
- Water scarcity affects more than 40% of the world's population and is expected to worsen in the future. Almost 1.7 billion people live in river basins where water usage exceeds recharge.
- In 80% of households without access to on-premises water, women and girls are responsible for collecting water.
- Floods and other water-related disasters account for 70% of all deaths related to natural disasters.
- At least 892 million people continue to practice open defecation.
- 2.4 billion people lack access to basic sanitation services, such as toilets or latrines.
- More than 80% of the wastewater generated by human activities is dumped into rivers or the sea without being treated for contamination.
- Each day, nearly 1000 children die due to preventable water and sanitation-related diarrhoeal diseases.

Although drinking water forms a very small part of the entire water requirement, it affects humans most in disease burden. Hence, ensuring adequate quantity and good quality remains the major challenge in achieving SDG 6. With the same water sources being used for various purposes, drinking water should be prioritised as the first necessity. The sources of water for meeting drinking water demand are given below.

1.2 SOURCES OF WATER

Water sources refer to water bodies that contribute to public drinking water and private wells. Meteorological water, such as rain, snow, hail, and sleet, is precipitated on the Earth's surface and can be considered the source of available water. Surface water and groundwater are the most common drinking

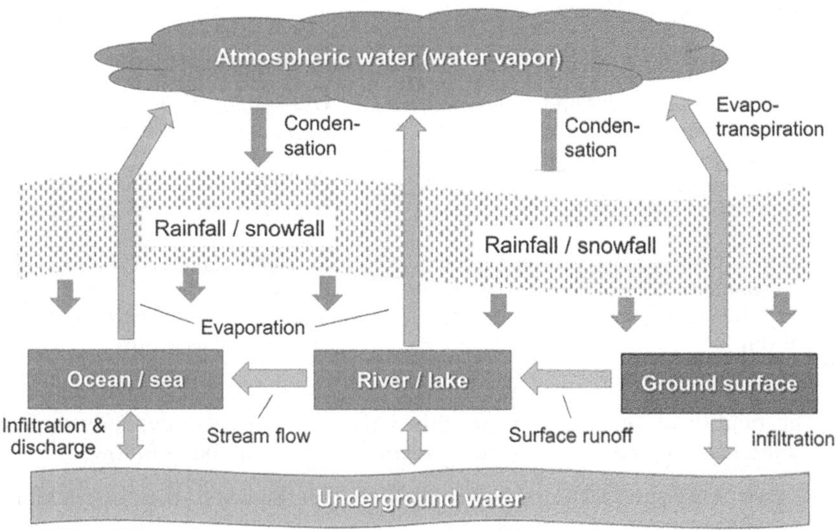

Figure 1.1 Sources of water (*source*: Wang & Luo, 2021).

water sources (Figure 1.1). Water is obtained by the community mainly from two sources, that is, surface water and groundwater.

1.2.1 Surface water

Surface water accumulation is primarily due to direct runoff from rain and snow. The proportion of rain and snow water that neither infiltrates into the ground nor returns to the atmosphere via evaporation and flows over the Earth's surface is considered direct runoff. Direct runoff is a major contributor to streams and ultimately to natural or constructed storage sites (or into the ocean in coastal areas). Considerable seasonal variation in surface water quantity is observed, mainly depending on rainfall. Surface water sources may further be classified as rivers, lakes, ponds and reservoirs created to store (artificial storage) water. Surface water can be transported to water treatment plants through the canal (open channel) before being distributed to households. Water in the ocean is also classified as surface water.

1.2.2 Groundwater

Water that infiltrates below the root zone eventually reaches a point when it fills all voids/crevices beneath the Earth's surface. The water in the saturation zone below the Earth's surface is groundwater, which can be extracted under the ground using a well that penetrates the water table. Dug wells with or without steining walls, dug cum bore wells, cavity bores, radial collector wells, infiltration galleries, tube wells, and bore wells are the most common methods for extracting groundwater.

Table 1.1 Activities responsible for water contamination.

Natural Factors	Anthropogenic Activities
• Weathering of bedrocks and geogenic contamination	• Over-exploitation
• Atmospheric process involving evapo-transpiration	• Agricultural runoff and change in irrigation practices
• Deposition of dust, salt and air pollutants by wind	• Discharge of untreated/partially treated domestic, municipal and industrial wastewater
• Natural disasters like floods and draught	• Eutrophication
• Inter-aquifer mixing of cold and hot water	• Salinity
• Infiltration through vegetation, swamps, and soil	• Municipal and industrial solid and hazardous wastes

The quality of surface and groundwater sources largely depends on the following factors, and the likely constituents of water can be as follows:

- Ca^{2+}, Mg^{2+}, and HCO^{3-} in groundwater from the water–rock interactions (major source).
- Atmospheric deposition of NO^{3-} (minor source).
- Na^+ and K^+ in groundwater from granite rocks (orthoclase and muscovite minerals).
- Cl^- from natural sources such as rain, fluid inclusion dissolution, and chloride-bearing minerals.
- Disturbance of riparian vegetation resulting in increased sedimentation of rivers (turbidity).
- Fluoride, arsenic, and other metals from rocks and soils and leaching into groundwater.
- Microorganisms and pathogens due to agricultural runoff, livestock grazing, mine drainage, urban runoff, and household and industrial discharges.
- Suspended solids from agricultural runoff and urban runoff, clay-rich mountain waters, leaching of soil contamination, and point source water pollution discharge from industrial or sewage treatment plants.
- Climate: pattern, strength, and duration of precipitation, as well as the temperature, intensity, and direction of wind movements.

The quality of surface and groundwater sources is affected by various activities broadly categorised into two classes – natural and anthropogenic. Table 1.1 presents various natural and man-made activities responsible for water contamination.

1.3 DRINKING WATER QUALITY PARAMETERS

Water is an excellent solvent. Hence, it assures the solubility of chemicals from natural and man-made sources, as evidenced by changes in indicator parameters like pH and electrical conductivity. There can be numerous constituents in water that can adversely affect water quality. The first step in ensuring safe

Table 1.2 Select water quality parameters.

Physical	Chemical	Microbiological	Emerging Contaminants
pH	Electrical conductivity	Bacteria	Pesticides
Colour	Heavy metals	Viruses	PPCPs
Taste and odour	Hardness	Protozoa	Radioactive elements
Turbidity/total suspended solids	Salinity	Helminths	
Total dissolved solids	Chloride, sulphate, nitrate	Phytoplankton	

drinking water is determining the characteristics to avoid adverse effects on consumption. Although all the water quality parameters do not adversely affect health, their determination has several applications ranging from aesthetic value to water treatment. Water quality is generally categorised into physical, chemical, and microbiological parameters (Table 1.2).

1.3.1 Physical parameters

1.3.1.1 pH

The pH is one of the most important water quality parameters. pH is defined as the inverse of the negative logarithm of hydrogen ion concentration. The carbon dioxide–bicarbonate–carbonate equilibrium controls the pH of water. An increased carbon dioxide concentration lowers the pH of the water, whereas a decrease in carbon dioxide concentration causes it to rise. The logarithmic pH scale ranges from 0 (very acidic) to 14 (extremely alkaline). The scale for pH and pOH is given in Figure 1.2.

The concentrations of substances in water can transition to a more toxic state as the pH changes. Change in pH affects ammonia toxicity, chlorine

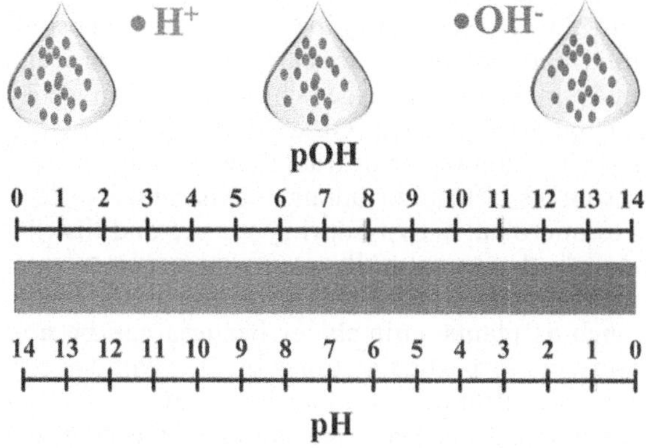

Figure 1.2 pH and pOH range.

disinfection capacity, pipeline corrosion, and metal solubility. Irritation of the eyes, skin, and mucous membranes occurs when pH levels are too high. Furthermore, because pH can influence the degree of metal corrosion and disinfection efficiency, it indirectly impacts health. As per WHO, pH is not of health concern at levels found in drinking water. Hence, no guideline value is proposed by WHO.

There are two methods of determining the pH. The first is calorimetric methods that use some indicator solutions or sometimes papers. Another is the electrochemical method which uses electrodes and a millivolt meter. This instrument is known as a pH meter.

1.3.1.2 Colour, taste and odour

Humans sense water quality based on colour, taste and odour. The colour of the water is mainly an aesthetic concern for checking water quality. The taste and odour of drinking water may indicate contamination or a problem with water treatment or distribution.

The presence of organic matter in water is indicated by its colour (mainly humic and fulvic acids). Colour contamination of water bodies is caused by dye pollution, soil particles, metals, and sometimes water blooms. The pigments of Cyanobacteria and Chlorophyceae colour the water when they form a bloom. The water turns brown when Bacillariophyceae has created a bloom. Flagellate Peridiuium occasionally blooms in dam reservoirs, turning the reddish water brown. Iron and other metals, either natural impurities or corrosive materials, also significantly impact colour. Colour is an important parameter because most water users prefer colourless water. Colour verification can aid in estimating the costs associated with water discolouration.

The taste of water may provide a clue about the type of contaminant present; salty (sodium chloride), sour (hydrochloric acid), sweet (sucrose), and bitter are all flavours that humans perceive. Relatively simple compounds produce sour and salty flavours. Sweet and bitter tastes, on the other hand, are formed by complex organic substances. The aroma and taste of water can be influenced by microbial, chemical, and physical elements.

Odour is created as a result of organic matter decomposition. Apart from this, several water treatment techniques can also cause unacceptable taste and odour; for example, improperly managed chlorination can form trichloramine, which has a foul odour and taste. Other issues, such as the disruption of internal pipe deposits, biofilms or degradation in water distribution systems, may be indirect sources of taste or odour. Furthermore, taste and odour can deteriorate due to microbial activity during storage and distribution. Sulphur or rotten egg odour is due to naturally occurring hydrogen sulphide. Bacteria can generate mouldy, musty, earthy, grassy, or fishy odours in a sink drain, or organic debris such as plants, animals, or bacteria can be present naturally in lakes and reservoirs at particular times of the year. The water may taste salty due to high concentrations of naturally occurring sodium, magnesium, or potassium. Such water taste may be found in the coastal region, where seawaters are supplied as freshwater after some water treatment.

There is no adverse health effect directly due to colour, taste or odour in water. However, consumers can switch over to unsafe water sources if water is found to have colour, obnoxious taste or odour.

Nessler tubes of 50 mL and Platinum-Cobalt colour (Pt-Co) are used to measure the colour of water samples. Sensory techniques examine a taste or odour's qualitative description and rate its intensity on a scale. The sensitivity of sensory techniques is generally higher than that of analytical instrumentation-based techniques. The taste threshold test, the threshold odour number (TON), the taste rating scale, and taste profile analysis are three standard procedures for examining taste and odour. TON entails diluting the water sample multiple times with reference water and comparing each dilution to the reference water. The TON is the greatest dilution at which odour may be detected. A TON of less than or equal to 3 meets the secondary drinking water criteria in the United States and Europe.

1.3.1.3 Turbidity/total suspended solids

After rainfalls, brown turbidity appears in rivers, reservoirs, lakes, and so on, resulting from rainwater carrying suspended solids into them. Particles larger than 2 mm in size are referred to total suspended solids (TSS). TSS may be sand, silt, plankton, and anything else that floats or 'suspends' in water. Although inorganic components make up the bulk of total suspended solids, algae and planktons are also included. Suspended solids include erosion and runoff, sediment disruption, sewage, industrial wastewater, algae, and so on. Turbidity/TSS alters the dissolved oxygen in the water, which hinders photosynthesis activity. Turbidity decreases the clarity of water to transmit light. TSS concentrations may also increase due to organic particles produced by decaying objects. When turbid water is obtained from a source, it affects the treatment plant by plugging treatment units. Increased turbidity in treated water indicates inefficient treatment units, for example filtration. In contrast, increased turbidity in the distribution system implies sloughing of biofilms and oxide scales and ingress of turbid water through faults such as main breaks. Turbidity reduces the aesthetic quality of the water streams. Moreover, it is harmful to recreation and tourism. High turbidity in water also increases the cost of water treatments. Turbidity of water is caused by various factors, including source water laden with suspended solids and the ingress of water through main breaks and other faults within distribution systems.

Nephelometric turbidity units (NTU) are used to evaluate the cloudiness of water caused by suspended particles (clay and silts), chemical precipitates (manganese and iron), and organic particles (plant waste), and species (NTU). This can be done by estimating the light intensity of the transmitted beam or measuring the light dispersed sideways. An electronic turbidity meter or a turbidity tube can measure turbidity. Turbidity does not have a health-based recommended value. However, the median turbidity must be less than 0.1 NTU for effective disinfection.

TSS can be accurately measured by filtration through a 0.45 mm membrane filter. Using a TSS sensor or monitor, one can correctly quantify total suspended solids in water. WHO does not specify guideline values for total suspended solids. A turbidimeter comprising a nephelometer with a light source utilised the principle in which the scattered light intensity by the sample under particular conditions is compared to the intensity of light scattered by a standard reference sample under identical conditions. Sample tubes (clear colourless glass) utilised the interference method. The scattered light intensity by the sample is compared to the intensity of light scattered by a standard reference sample under the same conditions.

1.3.1.4 Total dissolved solids

Total dissolved solids (TDS) is the most common ingredient in the form of cations such as calcium, magnesium, potassium, and sodium, as well as anions such as carbonates, hydrogen carbonate, chloride, sulphate, and nitrate. TDS of water is the residue left after evaporation and drying of a constant weight at temperatures varying from 103 to 105°C. TDS levels of water vary greatly across geographical regions due to variations in mineral solubility. TDS in drinking water covers many sources, including sewage, runoff from urban and agricultural land, and industrial wastewater. Natural sources of TDS include dissolution from the parent rock, mineral springs, carbonate deposits, and seawater intrusion. If the TDS of the water is too low, then it corrodes the metal piping and fixtures and therefore has a bitter or off-taste because of corrosion by-products. If the TDS level in the water is exceptionally high, the water will have a salty flavour, will corrode metal pipes, and will cause equipment to break prematurely.

The TDS does not indicate any specific health risk, but it can be used to provide insight into the status of the water over time and as a warning sign of a potential problem. WHO does not provide any guideline value for TDS.

Gravimetric analysis and conductivity are the two most common methods for determining total dissolved solids. While gravimetric analysis is precise, it is a time-consuming approach. Gravimetric analysis determines the weight of dissolved particles by weighing the solids that remain after the water has evaporated from the sample water. The capacity of a fluid to conduct an electric charge is measured in conductivity. Hence, most commonly used method of determining TDS in water is the measurement of specific conductivity. TDS levels are calculated from conductivity measurements using a factor that varies depending on water types. Using this method, TDS in water can be quantified to a practical limit of 10 mg L^{-1}. Individual constituents of TDS can also be measured.

1.3.1.5 Dissolved gases

Water, while transiting through the atmosphere and catchment, absorbs various gases such as methane, hydrogen sulphide, and carbon dioxide, depending on the sources of these gases, particularly in the catchment. Although absorption of these gases has a limited effect on these gases' physicochemical characteristics, removal of these gases from water is a prerequisite to properly maintaining water quality.

1.3.1.5.1 Carbon dioxide

The carbon dioxide (CO_2) level in surface waters is typically modest, ranging from 0 to 2 mg L^{-1}. Because of microscopic faunal respiration and lack of significant plant development near the lake/reservoir bottom, water can have a high CO_2 level. CO_2 concentration in groundwater varies considerably but is frequently higher than surface water. Carbon dioxide levels in the water that are too high (over 5–15 mg L^{-1}) might cause the following issues (Sala-Garrido *et al.*, 2021):

- Water becomes acidic (carbonic acid-H_2CO_3), making it corrosive.
- Iron remains in dissolved form, making it more difficult to remove.
- When lime is used to soften water, it reacts with CO_2, increasing the time required for the water-softening reaction.

1.3.1.5.2 Hydrogen sulphide

Hydrogen sulphide (H_2S), mostly found in groundwater, may be formed due to iron or sulphur-reducing bacteria in the well. H_2S is responsible for the rotten-egg odour sometimes found in groundwater. When even modest levels of H_2S are present in the water, the following severe operational issues arise:

- Chlorine for disinfecting the water becomes less effective.
- Corrosion of the different elements of the water supply system, such as pipes, water tanks, and so on.

1.3.1.5.3 Methane

Methane (CH_4) is found in the aquifers surrounding natural gas deposits. The garlic-like taste may be present in the water due to CH_4. Surface water sources receiving sewage may also have dissolved CH_4. CH_4 is very mildly soluble in water, has a low boiling point, and vaporises easily.

1.3.2 Chemical parameters

1.3.2.1 Hardness

Di-metallic cations such as calcium and magnesium ions cause water hardness. Temporary (carbonate) and permanent (non-carbonate) hardness are the two types of hardness. Hardness can be further divided into soft, moderate, hard, very hard, and extremely hard, as presented in Table 1.3. Boiling water can

Table 1.3 Classification of the hardness of water.

Hardness	Measure
Soft	0–100 mg L^{-1} as $CaCO_3$
Moderate	100–200 mg L^{-1} as $CaCO_3$
Hard	200–300 mg L^{-1} as $CaCO_3$
Very hard	300–500 mg L^{-1} as $CaCO_3$
Extremely hard	500–1000 mg L^{-1} as $CaCO_3$

Source: Omer (2019).

precipitate the temporary hardness. As per WHO guidelines, hardness is not of health concern at levels found in drinking water but may affect the acceptability. Hardness in water can be determined titrimetrically. Total hardness (mg L^{-1} as $CaCO_3$) = calcium hardness as $CaCO_3$ + magnesium hardness as $CaCO_3$.

Depending on the interaction of other parameters such as pH and alkalinity, water with a hardness greater than 200 mg L^{-1} might cause scale deposition in the treatment plant, distribution system, pipeline, and tanks inside buildings. When hard water is heated, calcium carbonate scale forms, soft water with a hardness of less than 100 mg L^{-1} has a low buffering ability, making it more corrosive to pipes.

1.3.2.2 *Major ions*

Several trace ions in water affect the chemical features of natural water and thus its potability. Ions make up the majority of dissolved inorganic compounds in freshwater. Atmospheric deposition, rock weathering, runoff, and other activities all contribute to the influx of these ions into water sources.

1.3.2.2.1 Sodium and potassium

Drinking water sources in most supplies have less than 20 mg L^{-1} of sodium (Mohsin *et al.*, 2013). Sources of sodium in water include saltwater intrusion, minerals, sewage and industrial discharges. Sodium is not acutely toxic at normal levels. However, acute effects are documented due to accidental overdoses. Excessive sodium intake can aggravate chronic heart disorders. WHO (2022) concluded no possible relationship between high sodium in drinking water and hypertension. Hence, health-based guideline value is not proposed by WHO. Sodium affects the taste of water above about 200 mg L^{-1}. A direct aspiration atomic absorption spectrophotometer can determine sodium.

Potassium concentration in natural freshwaters is typically low (<10 mg L^{-1}). Potassium ions are highly soluble and are required to survive most living things. Potassium levels in drinking water are typically modest and do not constitute a health risk. Potassium and salt are electrolytes that maintain human body fluid and blood volume resulting in proper functioning. WHO (2022) reported no evidence that potassium in treated drinking water is likely to pose a health risk. Hence, no health-based guideline value is established for potassium in drinking water.

Potassium can also be determined by direct aspiration atomic absorption spectrophotometry, flame atomic absorption spectrometry and inductively coupled plasma atomic emission spectrometry (ICP-AES).

1.3.2.2.2 Chloride

Many natural and anthropogenic sources, such as seawater intrusion, application of inorganic fertiliser, leachate from municipal waste landfill sites, sewage/industrial wastewater discharges and so on, contribute to water. Chloride in natural water without exposure to contamination is less than 10 mg L^{-1}. Toxicity due to chloride is observed in humans except in the rare case of impaired sodium chloride metabolism. High chloride in water raises electrical conductivity and consequently increases corrosion potential. Chloride can react with pipe material and form soluble salts resulting in metals in drinking water. Chloride above 250 mg L^{-1} can impart taste to water. WHO has not proposed any health-based guideline value for chloride in drinking water (WHO, 2022).

Chloride in water can be determined by several analytical techniques such as titration of silver nitrate with chromate indicator, automated iron (III) mercury (II) thiocyanate colorimetry, ion-selective electrode, silver colorimetry, and ion chromatography (IC).

1.3.2.2.3 Sulphate

Sulphates are naturally contributed through dissolution through several minerals such as barite ($BaSO_4$), epsomite ($MgSO_4 \cdot 7H_2O$) and gypsum ($CaSO_4 \cdot 2H_2O$). Sulphates can also be contributed through anthropogenic sources such as mining activities and wastewater discharges from pulp and paper, textile industries and tanneries beside sewage. A very high concentration of sulphate (>750 mg L^{-1}) is likely to have laxative effects. However, commonly observed sulphate in drinking water is not likely to have any adverse health effects. Hence, the WHO did not propose any guideline value for sulphate in drinking water (WHO, 2022). Sulphate is measured spectrophotometrically by adding barium chloranilate, which liberates chloranilic acid after reacting with sulphate ions.

1.3.2.2.4 Nitrate

Nitrate, an important plant nutrient, is present naturally in the environment. It enters both surface and groundwater due to agricultural activity, waste and sewage disposal, and bacteriological oxidation of nitrogenous materials in the soil. Nitrate varies considerably, but groundwater levels are often higher than surface water. In agricultural areas, nitrogen levels increased in water resources largely due to inorganic fertiliser and animal manure. Nitrate ion-selective electrodes are also available for nitrate determination in water. Methaemoglobinemia or blue-baby syndrome among infants is the major health outcome of nitrate in drinking water. Nitrite is also reported to react with compounds in the human gut to form N-nitroso compounds. Some N-nitroso compounds are carcinogenic in animals and are suspected to be carcinogenic in humans.

However, these data related to human carcinogenicity are only suggestive and not conclusive based on several epidemiological studies. The drinking-water nitrate exposure can alter human thyroid gland function, resulting in altered thyroid hormone concentrations and function. WHO has set the guideline value for nitrate in drinking water as 50 mg L^{-1} as nitrate ion (WHO, 2022). This nitrate level is protective against methaemoglobinemia and thyroid effects in the most vulnerable subpopulation, bottle-fed babies, and thus other population subgroups. Spectrophotometric techniques are used to determine nitrate in water, and detection limits range from 0.01 to 1 mg L^{-1}.

1.3.2.2.5 Fluoride

Fluorine is a natural element found in various minerals, including fluorspar, cryolite, and fluorapatite. Fluoride is naturally found in groundwater due to soil weathering and leaching runoff from fluoride-containing rocks. Fluoride in water can also be due to industrial wastewater discharge from aluminium smelters, and so on. Fluorides can cause fluorosis, which damages the teeth and bones, typically found in drinking water. Ingestion of moderate doses has dental consequences, but ingestion of high amounts over time might cause serious bone disorders. Excess fluoride can damage the parathyroid gland. This can result in hyperparathyroidism, which involves the uncontrolled secretion of parathyroid hormones. An excess amount of fluoride in the drinking water may also cause the health effects like acne and other skin problems, high blood pressure, myocardial damage, seizures and muscle spasms, high blood pressure and heart failure, and so on. When drinking water contains $3\text{–}6\,\text{mg L}^{-1}$ of fluoride, skeletal fluorosis (abnormal changes in bone structure) can occur. Excessive fluoride levels in drinking water can damage tooth enamel and cause mild dental fluorosis. Fluoride is typically determined using an ion-selective electrode, allowing the total volume of free and complex-bound fluoride dissolved in water to be measured. Fluoride can also be determined using a spectrophotometer. The guideline value of fluoride in water is $1.5\,\text{mg L}^{-1}$ as per WHO. Fluoride can be determined by using spectrophotometry and ion-selective electrodes (WHO, 2022).

1.3.2.3 Heavy metals

Heavy metal refers to any metallic chemical element with a relatively high density that is toxic when present in high concentrations in water. Among the heavy metals, chromium (Cr), mercury (Hg), arsenic (As), nickel (Ni), copper (Cu), cobalt (Co), cadmium (Cd), selenium (Se) and lead (Pb) are of major health concerns (high blood pressure, kidney malfunctioning, liver damage, stomach cramps, and diarrhoea, etc.). These are also natural components of the geological origin. Water flowing through soil and rock may dissolve iron and manganese minerals, retaining them in different amounts in solution. Iron and manganese levels are also affected by water quality factors such as pH, dissolved oxygen, and salinity. Iron and manganese in the dissolved form are also observed in deep surface water reservoirs with a depth of more than 20 m due to anoxic conditions. This is due to the accumulation of iron and manganese-containing sediments at the bottom of the reservoir having low dissolved oxygen. Due to prevailing anoxic conditions at the bottom of the deep reservoir, iron and manganese remain reduced and dissolved. Corroded pipes can also be a source of iron in drinking water.

Cadmium poisoning can harm the kidneys and cause skeletal damage over time. It is also a human carcinogen. Mercury poisoning damages the lungs if exposed to it for a short period. Anxiety, tremor, personality changes, restlessness, sleep disturbance, and depression are physiological and psychological signs of chronic mercury poisoning. Mercury, in its metallic form, can cause health problems. Contact eczema can be triggered by metallic mercury, and amalgam fillings can cause oral lichen. Headaches, stomach pain, irritability, and other nervous system symptoms are among the symptoms of acute lead poisoning. Inorganic arsenic is acutely poisonous, triggering gastrointestinal symptoms, serious cardiac and nervous system disorders, and sometimes death. Some side effects that survivors may suffer are bone marrow depression, haemolysis, hepatomegaly, melanosis, polyneuropathy, and encephalopathy. Intake of inorganic arsenic can cause peripheral vascular disease and gangrene. Sources, health effects and guideline values of various heavy metals are presented in Table 1.4. Figure 1.3 shows signs of arsenicosis and fluorosis. When water with iron and manganese is exposed to air, these elements oxidise and become less water-soluble. Coloured forms of iron and manganese appear in water after oxidation. Iron particles that are white, yellow, and eventually red-brown develop and drop out of the water. Iron oxide particles may not settle down and leave a red hue in the water (Sander *et al.*, 1996). Manganese that has

Table 1.4 Sources, adverse health effects and guideline values of heavy metals in water.

Element	Symbol	Sources	Adverse Health Effects	WHO Guideline Values (mg L^{-1})
Arsenic	As	Pesticides, chemical wastes, mining bi-products	Enzyme-inhibitor, carcinogenic	0.01
Cadmium	Cd	Industrial discharge, metal plating, Ni–Cd batteries, mining waste	High blood pressure, kidney malfunctioning, anaemia, bone marrow disorder	0.003
Chromium (III and VI)	Cr	Metal plating industries, tanning process.	Cr (VI) carcinogenic	0.05
Copper	Cu	Metal plating industries, mining, mineral leaching	Liver damage and kidney disease	2
Lead	Pb	Plumbing, mining, coal, gasoline	Causes anaemia, kidney malfunctioning, nervous disorder	0.0004
Manganese	Mn	Mining of industrial waste, microbial action of Mn minerals of low pH	Deficits in memory, attention, and motor skills, neurological	0.4
Molybdenum	Mo	Natural sources, industrial waste	Gout-like joint pain	0.07
Zinc	Zn	Metal plating industries, industrial waste	Vomiting, stomach cramps, and diarrhoea	3

Source: WHO (2022).

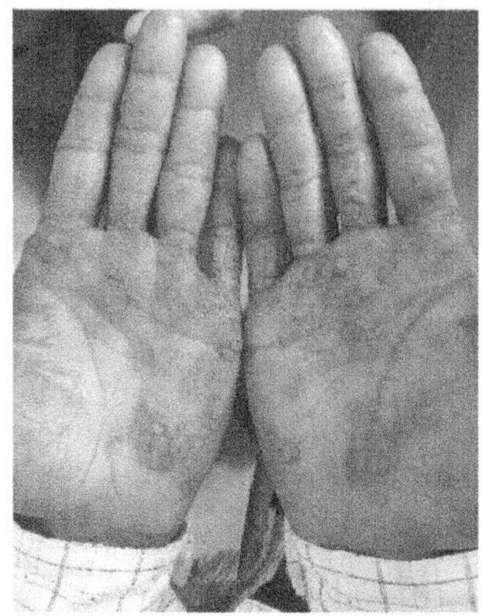
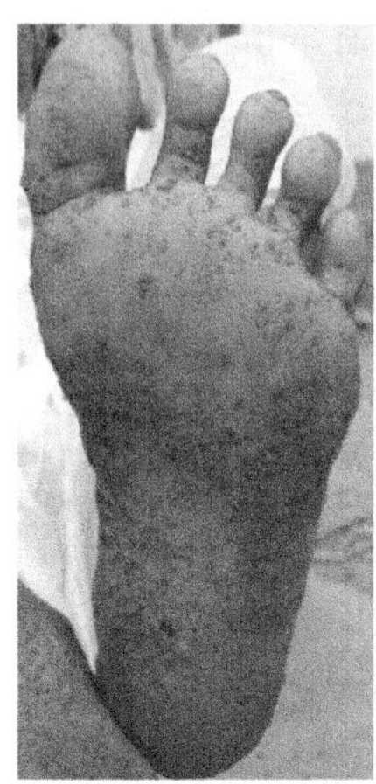

Arsenic poisoning

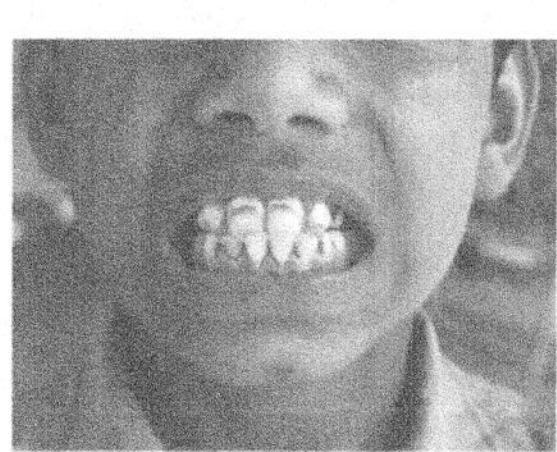
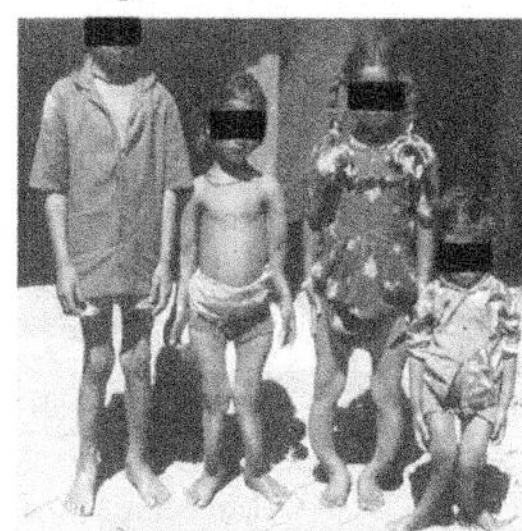

Fluoride poisoning

Figure 1.3 Health effects due to heavy metals in drinking water (courtesy of Dr Subhamoy Bhowmick CSIR-National Environmental Engineering Research Institute (NEERI), Kolkata, West Bengal, India and Dr Tapas Chakma, ICMR-National Institute for Research in Tribal Health (NIRTH), Jabalpur, Madhya Pradesh).

been oxidised generally remains dissolved in water, giving it a black colour. The staining qualities of water having high amounts of iron and manganese are caused by these sudden changes in the chemical forms of these elements. Water with more than $0.3\,\mathrm{mg\,L^{-1}}$ iron discolour plumbing fixtures and practically anything it comes in contact with. If the content is greater than $1\,\mathrm{mg\,L^{-1}}$, the water has a metallic taste and may be murky (Chaturvedi & Dave, 2012). Manganese in water causes blackish discolouration of fixtures even at levels as low as $0.1\,\mathrm{mg\,L^{-1}}$. Manganese concentrations of $0.1\,\mathrm{mg\,L^{-1}}$ and above can also cause staining issues (Lytle *et al.*, 2020).

Heavy metals are now determined using atomic absorption spectrophotometer, inductively coupled plasma-optical emission spectrometry (ICP-OES) and inductively coupled plasma-mass spectrometry (ICP-MS) at very low concentrations as guideline values/standards for these metals are also very low.

1.3.3 Microbiological

Microorganisms of concern in water are classified as bacteria, viruses, protozoa and helminths. Microbial contamination is one of the most important variables impacting water quality. Microbiological water quality can shift swiftly and significantly. Microbe monitoring results are rarely available to inform management and prevent polluted water from being distributed. Many microbiological contaminations are present in source water being used for drinking purposes. The presence of faecal coliform/*Escherichia coli* in water indicates contamination due to human or animal waste in the water. Phytoplankton is present mainly in surface water sources. Phytoplankton is classified into two groups: algae and cyanobacteria. Because of their potential to interfere with the water treatment process, which significantly affects water quality, two phytoplankton genera, *Cyanophyceae* and *Ceratium*, are identified as the problem-causing phytoplankton. *Cyanophyceae* blooms add a new dimension because toxic populations of various species can create cyanotoxins, which constitute a health hazard to water consumers. *Ceratium* disrupts coagulation, flocculation, and sedimentation processes, clogs sand filters, and creates taste and odour-producing chemicals.

In drinking water, pathogens that come from the stomachs of warm-blooded animals are the most dangerous. Animals in the wild, pets, and animals on farms or feedlots are all potential sources of contamination. Contamination problems arise from the improper dumping of raw sewage and wastewater treatment facilities that are wrongly designed, decaying, or overburdened, such as septic systems in private homes and leaking sanitary sewage lines. Anthropogenic sources are of particular concern since they contain bacteria derived from humans. While other sources of pathogen exposure may be vital, drinking water polluted with human and animal excreta poses the most significant risk to public health. Short-term peaks in pathogen concentration can considerably raise illness risks and potentially lead to waterborne disease outbreaks. Infectious disorders produced by pathogenic bacteria, viruses, protozoa and helminths are the most prevalent and pervasive health risks connected with drinking water. The principal transmitter of infections transferred through the faecal–oral pathway is drinking water. Faecal contamination is not the only source of microbial contamination in drinking water. Some species (such as *Legionella*) thrive in piped water systems, while others, such as the guinea worm (*Dracunculus medinensis*), thrive in source waters and can produce outbreaks and isolated instances. Examples include typhoid, cholera, infectious hepatitis (produced by the Hepatitis A or E viruses), and disease caused by *Shigella* spp. and *E. coli*. Others, like self-limiting diarrhoea, are usually associated with less serious outcomes (e.g., *Noroviruses, Cryptosporidium*). Even though many waterborne diseases have a low fatality rate, the socio-economic burden of even non-fatal infections is enormous.

1.3.3.1 Bacteria

Bacteria are single-cell microorganisms and are considered the first organism found on Earth. Bacteria can be found in water, soil, flora, fauna, and even radioactive waste. Bacteria can be beneficial and harmful and assist various life forms, including plants and animals. Almost one million bacteria are present in 1 ml of water. Bacteria are generally classified into five groups according to their shape: *Cocci* (spherical), *Bacilli* (rods), *Spirilla* (spiral), *Vibrio* (comma-shaped), and *spirochetes* (corkscrew). Bacteria can be found in fresh, brackish, or marine water.

In general, faecal coliform bacteria are the subgroup of coliform bacteria and are easy to detect compared to pathogens. They present in significant concentrations in human and animal intestines and faeces. Faecal coliforms, *E. coli* or thermotolerant coliforms are used as indicator bacteria. Faecal coliform or any other indicator bacteria in drinking water is a strong indicator of recent sewage or animal waste pollution. It should be taken as a sign that pathogens are present. Microbes in these wastes can produce short-term symptoms, including diarrhoea, cramps, nausea, headaches, other symptoms, and long-term health problems. Infants, small children, the elderly, and anyone with very weakened immune systems is at risk. However, pathogenic (disease-causing) bacteria in water can be harmful and potentially fatal for human health. For instance, pathogenic bacteria like *Salmonella*, *Shigella*, *Vibrio cholerae*, *Campylobacter*, *Clostridium*, or other enteric pathogens can cause infections such as typhoid, shigellosis, cholera, diarrhoea, botulism, or enteric fever and dysentery, respectively.

The major type of bacteria found in water is coliform bacteria. Coliform bacteria are rod-shaped, non-spore-forming bacteria found in the intestine of humans and animals. As a result, coliform bacteria are also present in the excretory matter. Therefore, the presence of coliform bacteria in water can indicate faecal contamination. Detection of contaminants like coliform bacteria is useful in monitoring water quality and preventing waterborne disease outbreaks. The most common detection methods are multiple tube fermentation or the membrane filtration technique. The multiple tube fermentation method uses the most probable number for the enumeration of bacteria, and the membrane filtration technique involves using a 0.45 μm membrane filter. As per WHO, *E. coli* or thermotolerant coliform bacteria in drinking water must not be detectable in any 100 ml sample. Health effects and routes of exposure to various bacteria are presented in Table 1.5.

1.3.3.2 Virus

Viruses contain different molecules consisting of genetic materials that need a host for survival. The genetic material can either be a single- or double-stranded deoxyribonucleic acid, or ribonucleic acid which has a protein coat. Viruses are microscopic organisms that can replicate and grow inside a host cell but are functionally inactive without the host cell. Viruses are a diverse group of organisms found in various habitats, including water. Much like bacteria, viruses enter the water through faecal contamination. However, unlike bacteria, viruses are not included in the normal microflora of the intestinal tract and are only excreted by infected individuals. However, the concentration excreted is usually several orders of magnitude lower than that of coliform

Table 1.5 Common bacteria that affect the management of drinking water supplies.

Microbial	Human Health Effects	Routes of Exposure
Acinetobacter	Urinary tract infections, pneumonia, bacteraemia, secondary meningitis, weakened immune system	
Aeromonas	Infections of the skin and the respiratory system	Soil and water-related activities (swimming and diving etc.)
Burkholderia pseudomallei	Melioidosis, pneumonia, internal organs and unusual neurological illnesses	Inhalation or ingestion, skin cuts or abrasions with contaminated water
Campylobacter	Abdominal pain, diarrhoea, vomiting, chills and fever	Food, consumption of animal products; particularly poultry products, and unpasteurised milk
Enterobacter sakazakii	Premature birth	Ingestion of non-sterile baby formula purchased from a store
Escherichia coli pathogenic strains	Diarrhoea, and fatal haemolytic uraemic syndrome	Person-to-person transmission, animal contact, contaminated food and water consumption.
Helicobacter pylori	Peptic, duodenal ulcer disease and gastric cancer.	Oral–oral transmission, person-to-person contact
Klebsiella	Immune systems damaging, and pneumonia	Contaminated water, aerosols in hospitals
Legionella	Legionellosis, Pontiac fever, kidney disease, cancer, diabetes	Devices for cooling using water (cooling towers and evaporative condensers, rivers, etc.)
Leptospira	Illness, muscle pain, chills, eye redness, abdominal discomfort, jaundice, and skin haemorrhages	Cuts and abrasions or via the mucous membranes of the mouth, direct contact with dead or living animals
Mycobacterium	The respiratory, gastrointestinal, and genitourinary tracts, as well as the skin and soft tissues	Inhalation, contact and ingestion of contaminated water.
Pseudomonas aeruginosa	Respiratory tract, physically damaged eyes.	Contamination of surgical tools or exposure of vulnerable tissue to contaminated water, such as wounds and mucous membranes.

(Continued)

Table 1.5 Common bacteria that affect the management of drinking water supplies (*Continued*).

Microbial	Human Health Effects	Routes of Exposure
Salmonella	Gastroenteritis, bacteraemia or septicaemia	Humans and a wide range of animals such as chickens, cows, pigs, sheep, birds, and even reptiles get infected.
Shigella	Bacillary dysentery, abdominal cramps, fever and watery diarrhoea	The faecal–oral pathway is used through person-to-person contact, contaminated food and drink.
Staphylococcus aureus	Boils, sepsis of the skin, post-operative wound infections, enteric infections, septicaemia, endocarditis, osteomyelitis, and pneumonia, as well as vomiting, diarrhoea, fever, abdominal cramps, electrolyte imbalance, and fluid loss	Hand contact, ham, poultry and potato and egg salads kept at room or higher temperature
Tsukamurella	Lung diseases, cutaneous and bone infections, meningitis and peritonitis.	The initial source of the contaminating organisms is unknown, as it is spread by devices such as catheters or lesions.
Vibrio	Illness	The faecal–oral route, contaminated water and food
Yersinia	Diarrhoea, fever and abdominal pain	Meat and meat products, milk and dairy products, and the faecal–oral pathway

Source: WHO (2022).

bacteria. Viruses in wastewater can be classified as having an outer envelope and no envelope.

The prevalence of enteric viruses in water poses a threat to public health. WHO classify several waterborne viral pathogens as having moderate-to-high health significance. Since viruses are parasitic and infectious agents, viruses in water can harm human health. Adenovirus, astrovirus, hepatitis A and E viruses, rotavirus, norovirus, other caliciviruses, and enteroviruses are common water-transmitted viral pathogens. Most of the viruses mentioned above are associated with gastroenteritis and diarrhoeal infections. However, it should be noted that viral pathogens can also cause severe illnesses like meningitis, encephalitis, hepatitis (hepatitis A and E viruses), myocarditis (enteroviruses), or even cancer (polyomavirus). The most commonly used molecular methods for detecting viruses in water are polymerase chain reaction (PCR) and quantitative real-time PCR (qPCR). Cell culture infectivity assays and immunoassays are the gold standard for detecting viruses. Viruses in water can also be detected by using microporous filters. Guideline values and limits of detection of various viruses are presented in Table 1.6.

Table 1.6 Common viruses that affect the management of drinking water supplies.

Microbial	Human Health Effects	Routes of Exposure
Adenoviruses	Respiratory diseases, pneumonia, pharyngoconjunctival fever	Person-to-person contact, hand-eye contact transmission
Astroviruses	Illness, gastroenteritis, predominantly diarrhoea	Faecal–oral route, person to person
Caliciviruses	Fever, chills, headache and muscular pain.	Contaminated food and water supplies, especially drinking water
Enteroviruses	Myocarditis, meningoencephalitis, poliomyelitis, herpes simplex, hand-foot-and-mouth disease, and newborn multi-organ failure are all diseases that can affect children.	Inhalation of airborne viruses, drinking water, and person-to-person contact
Hepatitis A virus	Damage to liver cells	Person-to-person spread, contaminated food and water
Hepatitis E virus	Hepatitis	Contacts and particularly nursing staff, person-to-person spread
Rotaviruses and orthoreovirus	Gastroenteritis	Oral–faecal route transmission of human rotaviruses from person to person and inhalation of airborne human rotaviruses

Source: WHO (2022).

1.3.3.3 Protozoa

Protozoa are unicellular, eukaryotic microorganisms and do not have cell walls. Protozoa can have various intracellular organelles dedicated to specific tasks. Protozoa can have very diverse morphological stages, depending on the species. Most protozoa in water have a cyst stage, which is highly resistant to external stress and dormant. In the disease-causing protozoan species, cysts are a common mode of infection, normally acquired through the faecal–oral route. Protozoa are unicellular eukaryotic organisms found in several different parts of the Earth. The most common water-related parasitic infections are cryptosporidiosis and giardiasis, caused by *Cryptosporidium* and *Giardia*. Sample collection and concentration, purification or separation of target organisms from other constituents in the sample, and a detection assay are the three principal stages in pathogenic protozoa detection and identification approaches in environmental samples. Colorimetric analysis, microscopy with dyes or fluorescent antibodies, molecular techniques (e.g., PCR), and infectivity in animals or tissue cultures are all examples of assay procedures. Health effects and routes of exposure of various protozoa are presented in Table 1.7.

1.3.3.4 Helminths

The term 'helminth' is derived from the Greek word 'worm'. It refers to all forms of worms, both free-living and parasitic. Helminth ova (eggs) will be present in the faeces of the infected individuals. The ova in the contaminated water enter the body, hatch, and become larval worms. The worms then may enter various organ systems, thereby causing an infection. All helminth detection techniques include general separation, recovery, and concentration steps. The helminths in water can be identified visually, and the helminthic ova can be identified by microscopic examination or filtration through a 10 μm cellulose acetate membrane or by Leeds I technique, including centrifugation, flotation, and ova recovery.

Pathogens spread by contaminated drinking water have diverse characteristics, behaviours, and resistance. Any pathogen in polluted drinking water can cause significant and even fatal sickness. Table 1.8 provides a general overview of pathogens that affect the management of drinking water supplies.

1.3.3.5 Phytoplankton

Phytoplankton are microscopic creatures that float in water. Although they are single-celled, they may form colonies big enough to be seen with the naked eye. Phytoplankton is photosynthetic, which means that they can transform carbon dioxide and water into energy using sunlight. Phytoplankton is classified into two groups: algae and cyanobacteria. Because of their potential to interfere with the water treatment process and significantly affect water quality, two phytoplankton genera, *Anabaena* and *Ceratium*, are identified as the problem-causing phytoplankton. *Ceratium*, for example, has been shown to disrupt coagulation, flocculation, and sedimentation processes, as well as clog sand filters and create taste and odour chemicals. *Cyanophyceae* blooms add a

Table 1.7 Common protozoa that affect the management of drinking water supplies.

Microbial	Human Health Effects	Routes of Exposure
Acanthamoeba	Granulomatous amoebic encephalitis	Swimming pools, surface water, and contact lens solutions
Balantidium coli	Diarrhoea, nausea, vomiting, headache and anorexia	Person to person, by contact with sick pigs, or through the ingestion of tainted water or food
Blastocystis	Diarrhoea, abdominal pain, anal itching, weight loss and excess gas.	The faecal–oral route
Cryptosporidium	Diarrhoea, sometimes including nausea, vomiting and fever	Faecal–oral route, contaminated food and water and direct contact with infected farm animals and possibly domestic pets.
Cyclospora cayetanensis	Watery diarrhoea, abdominal cramping, weight loss, anorexia, myalgia and occasionally vomiting	Faecal–oral route.
Entamoeba histolytica	Diarrhoea with cramping, lower abdominal pain, low-grade fever	Person-to-person contact, contamination of food
Giardia intestinalis	Diarrhoea and abdominal cramps	Person-to-person contact, contaminated drinking water, recreational water.
Isospora belli	Low-grade fever, mild diarrhoea and vague abdominal pain	Poor sanitation and faecally contaminated food and water
Microsporidia	Illness, kidney damage	Person-to-person contact and ingestion of spores in water or food contaminated with human faeces or urine
Naegleria fowleri		Swimming pools and spas, surface waters naturally heated by the sun
Toxoplasma gondii	Lymphadenopathy and hepatosplenomegaly	Contaminated soil or water

Source: WHO (2022).

new dimension because toxic populations of the various species can create cyanotoxins, which constitute a health hazard to potable water consumers. Water quality parameters and standard instrument requirements are presented in Table 1.9.

1.3.4 Emerging contaminants in water

The emerging contaminants are naturally occurring or anthropogenically generated chemicals and/or microorganisms that are rarely monitored in water. However, they can cause suspected or well-established adverse effects on health

Table 1.8 Pathogens that affect the manwagement of drinking water supplies.

Microbial	Human Health Effects	Routes of Exposure
Dracunculus medinensis	Erythema, dyspnoea, vomiting, pruritus and giddiness	Consumption of drinking water containing *Cyclops* spp.
Fasciola spp.	Vomiting, abdominal pain and a high fever up to 40°C	Drinking contaminated water
Free-living nematodes		Ingestion of drinking water, fresh vegetables fertilised with sewage
Schistosoma spp.	Fever, chills, muscle pains and cough	When people are exposed to free-swimming cercariae in contaminated water used for agricultural, household, or recreational activities, their skin is penetrated.

Source: WHO (2022).

or the environment. These emerging contaminants consist of pharmaceuticals and personal care products (PPCPs), surfactants, pesticides, insecticides, and industrial chemicals. Antibiotic-resistant bacteria are also emerging contaminants. The cause of concern is that the toxicological effects of these contaminants on the environment and humans are yet to be established, and many of these contaminants cannot be precisely tested in water as they exist in very small (nanogram or even in picogram) concentration. Apart from this, by-products of these contaminants are also formed when they pass through water treatment systems. These emerging contaminants are described below.

Table 1.9 Water quality parameters and instruments.

Water Quality Parameter	Instrument
Nitrate, phosphate and iron	UV–vis spectrophotometer
pH	pH meter
Turbidity	Turbidity meter
Conductivity	Electric conductivity meter
Sodium, potassium	Flame photometer
Anions and cations	IC
Fluoride	Ion-selective electrodes
Metals/heavy metals	ICP-MS
Pesticides, phenolic compounds, trihalomethane	Gas chromatography with FID/NPD/ECD detectors
Pesticides, PAHs, PCBs	Gas chromatography-tandem mass spectrometer
Pesticides, PGR and emerging contaminants	Liquid chromatography-tandem mass spectroscopy
Pesticides	HPLC

1.3.4.1 Pesticides

Pesticide/insecticide is often referred to any substance used to kill or control pests. In agriculture, herbicides (for weeds), insecticides (for insects), fungicides (for fungi), nematocides (for nematodes), and rodenticides are examples (vertebrate poisons). Chemical families can be used to classify insecticides. Organochlorines, organophosphates, and carbamates are three prominent insecticide families. Degradation is produced through the active ingredient's chemical, microbiological, or photochemical breakdown. Pesticide leaching pollutes groundwater. Leaching of the pesticide depends on the soil texture, pesticide characteristics, irrigation and rainfall.

The different pesticides in the water can be detected and determined with the reverse-phase high-performance liquid chromatography (HPLC), gas–liquid chromatography with electrolytic conductivity detection, gas chromatography (GC) with electron capture detector (ECD) after liquid–liquid extraction, UV detection, HPLC with UV absorbance detection, GC-mass spectrometry (MS), and so on. Guideline values and limits of detection of various pesticides are presented in Table 1.10.

1.3.4.2 Pharmaceutical and personal care products

PPCPs are products used by humans for personal health or cosmetic purposes, or agribusiness to improve livestock development or health. Pharmaceuticals may enter surface water and groundwater supplies by excretion from individuals using them, unregulated drug disposal (e.g., flushing drugs down toilets), and agricultural runoff from livestock manure. PPCPs generate physiological effects in humans at low concentrations. PPCPs get accumulated in the water through several pathways. Drugs are not completely broken down and absorbed by human bodies, so they are expelled and discharged into wastewater, entering the water supply.

Different disposal procedures, such as flushing down the toilet or draining in the sink, result in PPCPs ending in the water. Because of their ability to enter drinking water, they have become chemicals of growing public concern. Pharmaceutical compounds in water sources directly affect users, such as respiratory diseases, fertility problems, infections, chronic depression, and congenital problems such as mental retardation and physical anomalies. Because of their inherent potential to generate physiological effects in humans at low concentrations, PPCPs are a distinct category of emerging environmental pollutants. PPCPs get accumulated in the water through several pathways.

PPCPs in water samples are extracted using solid-phase extraction and determined using liquid chromatography coupled to the mass spectrometer.

1.3.4.3 Radioactive elements

Natural radioactivity exists in certain chemical elements in the atmosphere. Natural radionuclides, such as potassium-40, and radionuclides from the thorium and uranium decay sequence, such as radium-226, radium-228, uranium-234, uranium-238 and lead-210, are found in water due to natural processes (such as soil absorption) or artificial operations employing naturally

Table 1.10 Common pesticides that affect the management of drinking water supplies.

Pesticide	WHO Guideline Values (mg L^{-1})	Limit of Detection
Alachlor	0.02	0.1 mg L^{-1} by gas–liquid chromatography with electrolytic conductivity detection in the nitrogen mode or by capillary column GC with a nitrogen–phosphorus detector
Aldicarb	0.01	0.001 mg L^{-1} by reverse-phase HPLC with fluorescence detection
Aldrin and dieldrin	0.00003	0.003 mg L^{-1} for aldrin and 0.002 mg L^{-1} for dieldrin by GC with ECD
Atrazine and its metabolites	Atrazine and its chloro-s-triazine metabolites: 0.1, hydroxyatrazine: 0.2	Atrazine: 1 ng L^{-1}, isotope dilution MS with solid-phase extraction; 10 ng L^{-1}, GC-MS with solid-phase extraction; 50 ng L^{-1}, liquid chromatography (LC)–MS with solid-phase extraction; 100 ng L^{-1}, GC with nitrogen–phosphorus detection. Metabolites: 5 ng L^{-1}, capillary GC with nitrogen thermionic specific detection and HPLC with photodiode array absorption detection following extraction with styrene-divinylbenzene sorbents and elution with acetone
Carbofuran	0.007	
Chlorotoluron	0.03	0.1 μg L^{-1} by separation by reversed-phase HPLC followed by UV and electrochemical detection
Chlorpyrifos	0.03	1 μg L^{-1} by GC using ECD or flame photometric detection
Cyanazine	0.0006	0.01 μg L^{-1} by GC-MS
2,4-Dichlorophenoxybutyric acid	0.09	1 μg L^{-1} to 1 mg L^{-1} for various methods commonly used for the determination of chlorophenoxy herbicides in water, including solvent extraction, separation by GC, gas–liquid chromatography, thin-layer chromatography or HPLC, with ECD or UV detection
1,2-Dichloropropane	0.04	0.02 μg L^{-1} by purge-and-trap GC with an electrolytic conductivity detector or GC-MS
Dichlorprop	0.1	1 μg L^{-1} to 1 mg L^{-1} for various methods commonly used for the determination of chlorophenoxy herbicides in water, including solvent extraction, separation by GC, gas–liquid chromatography, thin-layer chromatography or HPLC, with ECD or UV detection

Table 1.10 Common pesticides that affect the management of drinking water supplies (*Continued*).

Pesticide	WHO Guideline Values (mg L^{-1})	Limit of Detection
Endrin	0.0006	0.002 µg L^{-1} by GC with ECD
Fenoprop	0.009	0.2 µg L^{-1} by either packed or capillary column GC with ECD
Hexachlorobutadiene	0.0006	0.01 µg L^{-1} by GC-MS; 0.18 µg L^{-1} by GC with ECD
Isoproturon	0.009	10–100 ng L^{-1} by reversed-phase HPLC followed by UV or ECD
Lindane	0.002	0.01 µg L^{-1} using GC
Mecoprop	0.01	0.01 µg L^{-1} by GC-MS; 0.01–0.02 µg L^{-1} by GC with ECD
Methoxychlor	0.02	0.001–0.01 µg L^{-1} by GC
Metolachlor	0.01	0.75–0.01 µg L^{-1} by GC with nitrogen–phosphorus detection
Molinate	0.006	0.01 µg L^{-1} by GC-MS
Pendimethalin	0.02	0.01 µg L^{-1} by GC-MS
Simazine	0.002	0.01 µg L^{-1} by GC-MS; 0.1–0.2 µg L^{-1} by GC with flame thermionic detection
2,4,5-Trichlorophenoxyacetic acid	0.009	0.02 µg L^{-1} by GC with ECD
Terbuthylazine	0.007	0.1 µg L^{-1} by HPLC with UV detection
Trifluralin	0.02	0.05 µg L^{-1} by GC with nitrogen–phosphorus detection

Source: WHO (2022).

Table 1.11 Common radioactive elements that affect the management of drinking water supplies.

Radioactive Elements	Guideline Values (Bq L^{-1})
Americium–241	1
Carbon-14	100
Uranium	0.03 (mg L^{-1})
Uranium-238	10
Uranium-234	1
Thorium-230	1
Radium-226	1
Radium-228	0.1
Lead-210	0.1
Thorium-228	1
Thorium-232	1
Polonium-210	0.1
Plutonium-239	1
Caesium-134	10
Caesium-137	10
Strontium-90	10
Tritium	10 000

Source: WHO (2022).

existing radioactive elements (e.g., mining and processing of mineral sands and pesticide production). Small amounts of radioactive radium and uranium can be found in practically every rock and soil and dissolve in water. Radon, a radioactive gas produced by the disintegration of radium, can also be found in groundwater naturally.

Radioactive compounds (radionuclides) can be found in drinking water and pose a health risk to humans. These contaminations in the drinking water may cause radiation damage, bone marrow fatality, cataract initiation, cancer stimulation, cholera, dysentery, tuberculosis and pneumonia. At doses above 100 mSv, evidence of elevated cancer risk in humans is known for repeated exposures, such as ingestion of radionuclide-contaminated drinking water over long periods. Uranium in drinking water causes cytotoxic damage to the kidney nephron's proximal tubule. In males, it also induces bone resorption. Natural uranium causes chemical toxicity, especially nephrotoxicity, which is more dangerous than radiotoxicity; radium and radon, on the other hand, are known to cause radiotoxicity. With the increased use of 226Ra, bladder cancer in men, breast cancer in women, and lung cancer in both sexes have increased (Canu *et al.*, 2011).

ICP-MS by solid fluorimetry with either laser excitation or UV light, ICP using adsorption with chelating resin. Guideline values of various radioactive elements are presented in Table 1.11.

1.4 SUMMARY

There is a reduction in water quantity, and quality is also adversely affected primarily due to anthropogenic activities. Surface water and groundwater are the most common sources of drinking water supply, although alternate sources such as seawater are also being developed to meet growing water demand. There can be numerous constituents in water that affect water quality. Hence, determining water quality parameters is the first step in ensuring safe drinking water. Determination of water quality has several applications ranging from aesthetic value to water treatment besides avoiding adverse health effects. These water quality parameters are broadly classified into physical, chemical, radiological, and microbiological, particularly from the consideration of PoU water treatment systems.

Most PoU water treatment systems are developed to remove pathogens from drinking water. However, membrane-based (RO) PoU water treatment systems can remove almost all these constituents from water. Microorganisms of concern in drinking water are categorised as bacteria, viruses, protozoa and helminths. Many microbiological contaminations are present in source water being used for drinking purposes. Hence, monitoring water quality, particularly microbiological water quality, throughout the water supply cycle is important and can complement the water safety plan. Various techniques are available for monitoring water quality parameters.

doi: 10.2166/9781789062724_0029

Chapter 2

Water supply systems and the need for point-of-use treatment systems

2.1 INTRODUCTION

The availability of safe drinking water is critical for human health everywhere. Due to limited availability, many water sources are used for numerous purposes such as drinking/domestic, irrigation, navigation, and intended/unintended waste disposal, resulting in quality degradation. Water is used for drinking, cooking, cleaning, removing trash, and other household purposes. Water supply systems must also meet the needs of public, commercial, and industrial users. Water must meet both quality and quantity standards in all these circumstances. Water cannot be used naturally because many constituents are beyond a specified limit, often called standards. The presence of organic, inorganic and microbiological constituents complicates producing drinking water. Concerns over water quality, particularly for drinking and household purposes, have led to water treatment technologies that improve water quality to varying degrees. A water treatment system aims to prevent acute diseases caused by pathogens, chemicals and micro pollutants, and finally to create palatable drinking water that does not deteriorate in quality during transportation from the treatment plant to the consumer. Conventional water treatment processes aim to eliminate contaminants such as suspended particles, natural organic matter, dissolved iron and manganese, and so on, naturally, besides microorganisms. However, the technologies' performance in removing micropollutants (chemicals) is limited.

In general, a water supply system consists of three major components: (1) source of supply, (2) treatment, and (3) distribution to the consumers. Water is transported to the treatment plant either by pressure (pumping) through the pipeline or open-channel (canal) flow. After the treatment, water, in most cases, is pumped to the distribution system directly with the network of elevated/ ground-based reservoirs. Over a century, there has not been any significant variation in water supply systems, and design criteria that evolved earlier are still being used. With the emergence of the water safety plan concept, the complexity of the water supply system is further reduced. A focus of the water safety plan is more on the prevention of water contamination; process units of conventional water treatment plants are reduced/bypassed/eliminated. However, there is growing emphasis on safe handling and treatment of water in households as ensuring safe water supply to the household through the system is an over-ambitious goal. This has led to the development and installation of point-of-use (PoU) water treatment systems.

This chapter gives an overview of water supply systems and the need for membrane-based PoU water treatment systems. In the PoU water treatment system, a single or chain (multi-barrier approach) of treatment processes is used to generate water of desired quality depending on the raw water quality.

2.2 WATER SUPPLY SYSTEMS

Water supply systems in their entirety include the following components that also affect water quality (Figure 2.1). This scheme is representative and can also represent surface water and groundwater sources.

- catchment (watershed)
- source
- treatment plant
- distribution network
- households

While water quality factors in the catchment largely remain unattended, particularly in developing countries, a typical water supply system comprises sources, treatment plants, distribution networks, and consumers (Medema *et al.*, 2003). The water treatment plant receives water from the surface water source through pipes under pressure or by open channel, while a tube/bore well remains the major conduit for groundwater sources. The source remains a major concern from a water quality perspective.

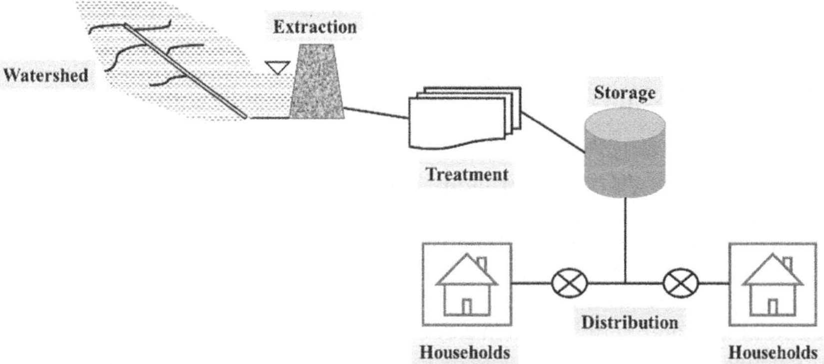

Figure 2.1 Components of the water supply system.

2.2.1 Catchment

Water supply engineers hardly control catchment while designing water supply schemes. However, the water quality of any surface water source is greatly influenced by catchment characteristics. Land use/land cover, developmental activities including agriculture practices, geology, soil and hydrology are the key characteristics of the catchment that affect water quality in terms of suspended solids, dissolved solids, nutrients and emerging contaminants such as pesticides. The relationship between land use/land cover of the catchment and surface water quality is complex. The linkage between anthropogenic factors due to land use/land cover changes on stream water quality is extensively studied (Figure 2.2).

The contribution of any water quality constituent is governed by prevalence, mobilisation and transportation to the surface water source. Constituents can be present naturally (suspended solids such as clay) or contributed externally, such as nutrients, pesticides and so on. Further, constituents are mobilised by various erosion, weathering, and desorption processes. Nutrients in dissolved forms are mobilised either through runoff or external causes. Solids are present in the dissolved forms, normally mobilised due to weathering and dissolution from the rock and soils besides external contribution (to the catchment). Mobilisation of constituents can also occur instream through bank erosion, decay of organic matter or nutrient cycling.

A high load of suspended solids can contribute to turbid water, which affects habitats of biota and decreases light penetration. Moreover, microorganisms, heavy metals and persistent organic pollutants can adhere to these sediments. Nitrogen and phosphorus are present in suspended and dissolved forms, leading to eutrophication, particularly in stagnant surface water sources

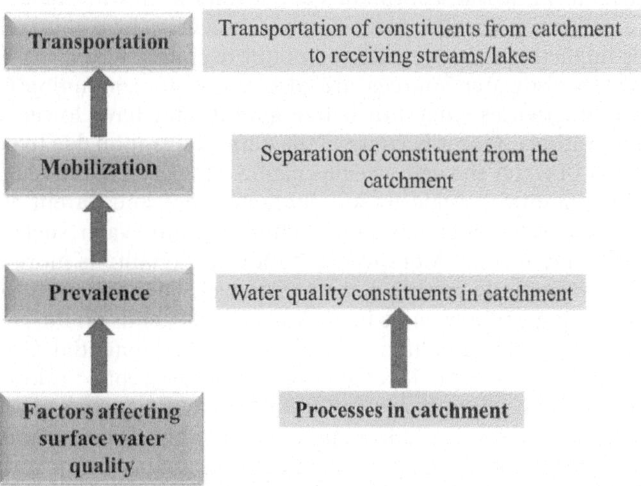

Figure 2.2 Process influencing the contaminants in streams and rivers.

such as reservoirs, ponds, and lakes. Transportation of constituents from the catchment is mainly driven by hydrological and meteorological parameters such as the quantity of rainfall and resultant surface runoff, slope, stream network and density, land use/land cover, soil characteristics and so on. An overground small stream network transports suspended solids into receiving water bodies, whereas subsurface flows transport finer solids.

2.2.2 Source

For a conventional public water supply system, the water source is primarily selected to provide sufficient quantity to meet projected water demand for a design period. Water demand considers all domestic, commercial, institutional, fire-fighting and, in some cases, industrial usages. Considering growing urban centres, multiple sources, surface water or groundwater, cater to ever-increasing demand. Surface water sources can be perennial rivers, reservoirs or lakes, whereas groundwater can be obtained through the tube/bore wells. Groundwater remains a major drinking water source for most rural people in developing countries despite surface water sources.

Improved water source that is likely to protect from source contamination is often recognised as one of the significant interventions for households' access to safe water. However, contamination during water collection, transportation, and household storage offsets this benefit.

Surface water sources are usually identified upstream of major urban centres to avoid contamination due to sewage. However, growing urban centres and the proximity of these centres restrict the selection of water sources only upstream of the cities. Protected surface water sources have inorganic and organic suspended solids, dissolved solids and microorganisms. As discharges from point and non-point domestic, agriculture and industrial sources are inevitable in large parts of the world, continuous deterioration in water quality of these sources is a major concern for water supply engineers. These sources also have compounds imparting colour, taste, and odour to source water.

Although groundwater sources are also prone to contamination due to anthropogenic activities, they are better quality and have lower suspended solids concentrations. This is because of the natural filtration due to percolation and movement of water through various media. However, dissolved solids may be higher in groundwater due to soil characteristics and parent rock. Some dissolved solids may be geogenic contaminants in groundwater, such as fluoride, arsenic, and heavy metals. Moreover, groundwater quality is more consistent temporarily and spatially (within a short distance). The concentration of these naturally occurring contaminants in groundwater depends on the geochemical conditions, such as pH and reduction–oxidation (redox) potential. Groundwater sources such as tube wells are located close to latrines, septic tanks and ponds, resulting in contamination due to infiltration in aquifers from these sources. Activities such as mining, agriculture practices, sewage disposal, and extensive groundwater extraction accelerate chemical contamination of groundwater. The groundwater quality is rapidly deteriorating, particularly in developing countries, primarily due to the over-extraction of these sources.

2.2.3 Treatment plant

Surface water sources are subjected to conventional water treatment, and the following chain of units is provided. Principal contaminants and corresponding treatment units of the conventional water treatment plant are shown in Figure 2.3. The primary objective of a conventional water treatment plant is to remove suspended solids (turbidity) and microorganisms (Zhang *et al.*, 2012). The water source is presumed to be protected, and a conventional water treatment plant is provided to remove naturally occurring contaminants. Other sources of contaminants are either properly treated before disposal, or sufficient dilution is available in recipient water bodies.

Conventional water treatment is challenging when source water quality deteriorates, resulting in higher treatment costs and reduced water quality in households. Surface waters and groundwater resources' corrosiveness depends upon pH, hardness, and other factors. Some water streams contain dissolved minerals, which accumulate inside pipelines and cause scale to form. In conjunction with other forms of treatment, highly corrosive raw waters can be treated to lessen this feature. The temperature of treated water is similar to that of raw water. During the detention duration in the treatment facility, the ambient air temperature may cause minor alterations. High water temperatures hasten corrosive action and reduce water viscosity.

If geogenic contaminants are not present, groundwater may require only disinfection before supply. If geogenic contaminants are present and no alternate water source is available, treatment specific to the contaminant is inevitable. Groundwater requires less extensive treatment than surface water. In addition to disinfection, raw water treatment may involve coagulation, sedimentation, filtration, softening, and iron removal. Water treatment technologies for geogenic contaminants can be classified based on their location at the source or the PoU. Community or household level treatment can be another classification. Water treatment technology can also be classified based on treatment principles (filtration, adsorption, electrocoagulation, etc.).

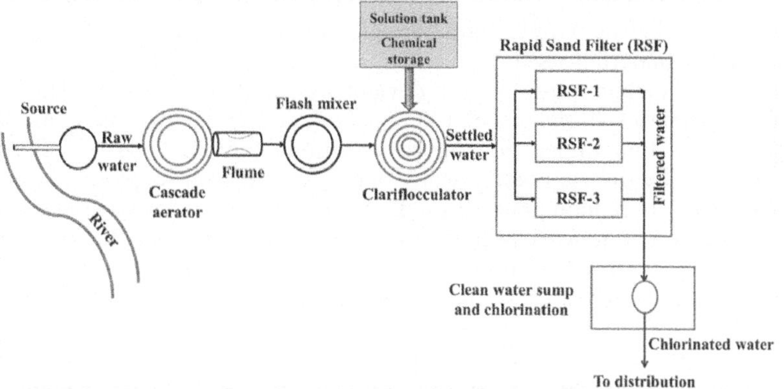

Figure 2.3 Major units of the conventional water treatment plant.

2.2.4 Distribution network

A distribution network or system is probably the most important and expensive element of the entire water supply system. The network aims to provide drinking water to consumers in sufficient quantity at an adequate pressure. Water is transported to and from the treatment plant through a pipe under pressure. A distribution network interconnects pipes, ground-based/elevated storage reservoirs, valves, service connections, intermediate pumping systems, fire hydrants, and household service connections (Cunha & Sousa, 1999). In addition, water is also supplied to commercial, industrial and institutional entities through the same distribution network. Although a minimum of 20 m of pressure is provided at the tail end in developed countries, achieving such high pressure is challenging in most developing countries primarily due to intermittent supply, rapidly growing networks, illegal connections and high energy costs. Mains and pipes in the distribution network are designed to withstand the design pressure. Flow in the distribution network is often controlled by gravity and pressure (pumping).

Water is stored at predefined elevated locations to provide sufficient pressure for transporting water to the tail end of the distribution network. However, in some instances, pumps are required to provide the designed pressure in the network. As the distribution network becomes more complex, the combined gravity and pumping system become inevitable. Water reservoirs are provided at intermediate locations, and it has now become a practice to provide these intermediate reservoirs separately for individual district metered areas. Chlorination is also being provided at reservoirs to maintain appropriate residual chlorine levels (0.2–0.5 mg L^{-1}) to protect against chlorine loss in the leaking distribution network.

Pipes for the distribution network come in a variety of lengths and sizes. Cast iron, ductile iron, plastic, and reinforced concrete are commonly used in mains. Galvanised iron, plastic, cast iron, or ductile iron pipe can be used for service lines (household connections). Linings are used to smooth out the inside of pipes and prevent corrosion. Polyvinyl chloride (PVC), polyethene (PE), acrylonitrile-butadiene-styrene (ABS), polybutylene (PB) plastic, high-density poly ethylene (HDPE), and fibre-glass-reinforced plastic (FRP) are some of the plastic materials used.

As water supply in most cities in developing countries is intermittent, water quality deteriorates in the distribution network largely due to cross-connection and leakages. Impounded/flowing water around the pipe gets into the pipe due to back-flow into a potable water supply during non-supply hours as the pipeline does not remain pressurised. Back-flow of water occurs as the pressure in the distribution network is due to lower pressure than atmospheric pressure. Water quality degradation in the distribution network is serious in developing countries due to inadequate resources for maintaining the network and chlorine residual. Rapid urbanisation is often associated with overwhelming demand and unplanned expansion of distribution networks. Intermittent water supply, low pressure, inadequate disinfection residual, leaks, and corrosion contribute to water quality degradation in the distribution network. Poor operations and maintenance, equipment failure, and external factors such as illegal connections affect water quality through distribution networks.

2.2.5 Households

Although the water systems can provide high-quality water, there is a significant risk of deterioration due to the amount of handling between the source of water and consumption. Water quality degradation through the supply chain has been extensively reported globally in water supply systems such as piped water, boreholes, handpump attached tube wells and even traditional sources. It may be claimed that unless careful consideration is given to how water is collected, stored, and managed, ineffective solutions will be presented, and resources will be wasted. Whether or if household water is of adequate microbiological purity at the outset, it is frequently contaminated with faecal germs during collection and storage due to unsanitary handling methods.

The following can be possible reasons for the deterioration of water quality in households which also include poor hygiene practices in locations having intermittent water supply:

- Type and material of containers for drinking-water storage.
- Washing hands prior to collection of water.
- Washing of container prior to collection of water.
- The lid over the water container.
- Technique for fetching water from the container (poured, ladled, dipped).
- Ladle handling practices.
- Household water treatment practice.
- Storage duration of water in the container.
- Hygienic conditions (e.g., access to animals etc.) around the water container.

2.3 MAJOR CONTAMINANTS IN WATER SOURCES

Although access to safe drinking water is one of the most important requirements for a healthy existence, the waterborne disease remains a leading cause of death in many regions, particularly among children, and a significant economic restriction in many subsistence economies.

Surface waters, such as rivers and reservoirs, and groundwater are the two main drinking water sources. Water contamination occurs as a result of both natural and man-made activities (Palansooriya *et al.*, 2020). Commonly observed water contamination sources are shown in Figure 2.4. Natural contaminants, particularly inorganic contaminants derived from the geological layers through which the water flows and anthropogenic pollution by microorganisms and chemicals, are present in water to variable degrees. Groundwater is less susceptible to pollution than surface waterways in general. Man-made pollutants can come from various places, some of which are more important than others. These can be classified as either point or diffuse sources. Industrial discharges and sewage treatment plants are point sources, making them easier to identify and control; runoff from agricultural land and hard surfaces, such as roadways, is less evident and difficult to

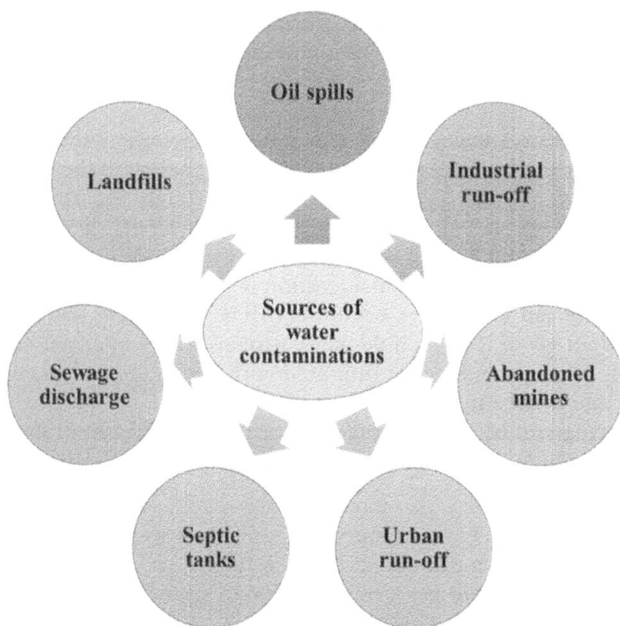

Figure 2.4 Sources of common water contamination.

control. As a result of these sources, the pollutant burden can change dramatically over time. There is the risk of chemical spills from industry and agriculture and pathogen-laden slurries from intensive farm units. In some nations, improperly placed latrines and septic tanks are a major cause of contamination, particularly for wells. Local industries can also contaminate water sources, particularly when chemicals are handled and disposed of improperly. Excessive growth of cyanobacteria or blue-green algae can result from fertiliser runoff or leaching into slow-moving or motionless surface waters (Yadav, 2006). Many species produce noxious compounds that can cause bad taste and odour and obstruct drinking water purification. However, they frequently release pollutants that are harmful to one's health, especially if treatment options are restricted.

Unwanted residues of chemicals used in water treatment might create contamination and sedimentation in water pipes if treatment is not improved. Contamination during water distribution can occur due to materials such as iron corroding and releasing iron oxides or pollutants entering the system. When oil spills into the surrounding soil, diffusion via plastic pipes might occur, causing issues of taste and aroma. Contamination can also occur in consumers' homes due to materials used in plumbing, such as lead or copper, or as a result of liquids back-flowing into the distribution system due to faulty connections. Chemical or microbiological pollutants are examples of such contaminants. The major water contaminants are discussed in Chapter 1.

2.4 TREATMENT FOR SURFACE WATER SOURCES

Conventional water treatment plants are primarily designed to remove contaminants present in natural surface water sources. As these conventional water treatment plants were designed and implemented many years back, this assumption of the protected catchment is no longer valid. Nonetheless, these plants remain the only option to treat surface water in many countries. A combination of processes, viz., aeration, coagulation, flocculation, sedimentation, filtration, and disinfection are used in the conventional water treatment plant. Each processing unit can be tuned to reach the required water quality after treatment during the design and operating stages, depending on the raw water quality. Rapid mixing of coagulants, flocculation and sedimentation removes suspended and colloidal solids in water. Filtration can remove colloidal and other finer solids and dissolve organic and microbiological contaminants. These processes can be carried out effectively using mechanical or hydraulic actions. Disinfection of water is performed as a final step in improving water quality by utilising chemicals or irradiation techniques to eliminate disease-causing microorganisms. Process units of the water treatment plant and contaminants removed in each unit are presented in Table 2.1.

There might be a slight deviation in water treatment plant units based on various factors such as quality of source water, the quantity of water to be treated, funds availability for construction, operation and maintenance of water treatment plant and so on. Details of various units of conventional water treatment plants are given in subsequent subsections.

2.4.1 Aerator

The aerator is often the first unit in any water treatment plant. The aerator brings water and air into close contact by releasing droplets or thin layers of water into the air or introducing tiny air bubbles. Contaminants are eliminated or modified during aeration. Aeration can remove dissolved gases like CO_2, CH_4, and H_2S. A few metals like iron and manganese are oxidised during aeration, resulting in the precipitated form that filtration can remove. The efficiency of the aeration

Table 2.1 Principal contaminants and corresponding water treatment unit.

Contaminants	Treatment Unit
Taste and odour-causing volatile organics and gases, iron and manganese	Aeration
Microorganisms and taste-/odour-/colour-causing compounds	Pre-oxidation (pre-chlorination)
Colloidal particles	Coagulation Flocculation
Suspended particles (solids) or floc produced in treatment processes	Sedimentation
Suspended particles (solids)	Filtration
Pathogens	Disinfection

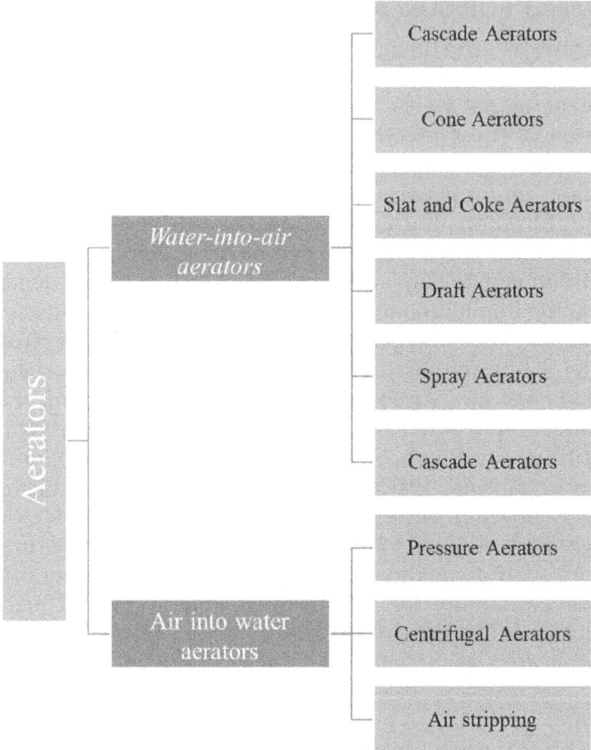

Figure 2.5 Types of aerators in the conventional water treatment plant.

process depends on the surface contact of air–water quantity, mostly governed by the size of the water droplets or air bubbles. Adding oxygen to contaminated water improves the water palatability as the bland taste is removed.

Aerators can be divided into two categories (Figure 2.5) based on the introduction of

(i) air to water,
(ii) water to air.

The water-in-air method involves injecting microscopic air bubbles into the water stream, whereas the air-in-water method involves creating small water drops that fall through the air.

A cascade aerator (the oldest and most frequently used) consists of a series of steps that the water flows over (similar to a flowing stream), as shown in Figure 2.6. In a cascade aerator, aeration takes place in the splash zones. Blocks are placed over the incline to form splash zones. Cascade aerators can oxidise iron and reduce dissolved gases to some extent.

Figure 2.6 Cascade aerator in water treatment plant (courtesy of Nagpur Municipal Corporation and Orange City Water Private Limited, Nagpur, India).

2.4.2 Pre-sedimentation

Pre-sedimentation (pre-sed) is an effective water treatment process commonly employed with high turbidity source waters. The pre-sed unit reduces suspended solids from raw water to more acceptable levels and improves the performance of subsequent treatment units. The pre-sed process can range from basic operations, such as settling ponds, in which hydraulic retention time is the important variable used to settle large solids, to more complex approaches, which may remove considerable turbidity and organics in the source waters. The objective of pre-sed is to reduce source water suspended solids, turbidity, and/or organics as easily and sustainably as possible before transporting the water into other treatment units. Figure 2.7 shows the photograph of the pre-sed tank in a water treatment plant.

When water is laden with a high suspended solids concentration, mechanical devices are usually utilised. A cyclone degritter is a centrifugal sand and grit removal equipment. As solid-laden water enters the unit, it spirals within the cylindrical part in a spiral route. The particles are thrown towards the cylinder wall by centrifugal force, which is often the force of gravity. The particles then spiral towards the cone's tiny end, dumping them into the sand accumulator tank and some water. The clean water exits the device through the vortex finder, nearly completely free of sand. A sedimentation (settling) tank can also be designed and constructed as a pre-sed unit to add redundancy to the water treatment plant. Three types of pre-sed systems are as follows:

- pre-sed impoundments,
- sand traps,
- mechanical sand-and-grit removal devices.

Figure 2.7 Pre-sedimentation tank in a water treatment plant (courtesy of Nagpur Municipal Corporation and Orange City Water Private Limited, Nagpur, India).

2.4.3 Coagulation, flocculation, and sedimentation

Coagulation and flocculation are often performed after a physical separation in water treatment followed by sedimentation (Malkoske *et al.*, 2020). Water turbidity (the cloudiness or haziness of a fluid, i.e. normally imperceptible to the human eye) is a fundamental indicator of water quality, and these three techniques give a proven process for treating it. In wastewater treatment, they can reduce suspended particles and organic loads by up to 90%. Coagulation, flocculation and sedimentation units are shown in Figure 2.8.

2.4.3.1 Coagulation

Two fundamental mechanisms cause coagulation: In electrokinetic coagulation, ions or colloids with opposite charges reduce the zeta potential to a level below the Van der Waals attractive forces, whereas, in orthokinetic coagulation, micelles combine and form clumps that agglomerate the colloidal particles (Sillanpää *et al.*, 2018). The majority of colloidal particles in water are stable, negatively charged, and repel one another. A high-energy, rapid mix is required to appropriately spread the coagulant and enhance particle collisions to produce good coagulation. The coagulant neutralises the charge allowing

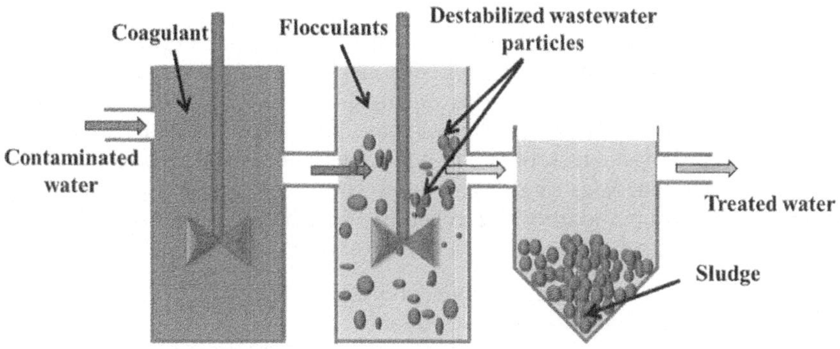

Figure 2.8 Coagulation, flocculation, and sedimentation units.

particles to come closer together and create big particles easily removed from the water. To neutralise the negative charges on non-settleable solids, coagulant chemicals with charges opposite suspended solids are introduced to the water (such as clay). The little dispersed particles might cling together once the charge has been neutralised. Coagulation is unaffected by excessive mixing, whereas insufficient mixing leaves this process unfinished. In a rapid-mix chamber, contact time is usually 1–3 min. Microflocs are somewhat larger particles that are not visible to the human eye.

2.4.3.1.1 Primary coagulants

Coagulants are either metallic salts (aluminium or ferric salts) or polymers. The ability of these coagulants to generate multi-charged polynuclear complexes in solution with improved adsorption characteristics is the main source of their efficacy. The pH of the system can influence the type of complexes that form. Polymers are organic compounds consisting of a lengthy chain of smaller molecules. Polymers are classified as cationic, anionic, or non-ionic. Metal coagulants are classified into two categories:

Aluminium-based: Aluminium sulphate, aluminium chloride, sodium aluminate, aluminium chlorohydrate, polyaluminium chloride (PAC), polyaluminium sulphate chloride, polyaluminium silicate chloride, and polyaluminium chloride-containing organic polymers are some of the aluminium coagulants.

Iron-based: Ferric sulphate, ferrous sulphate, ferric chloride, ferric chloride sulphate, poly ferric sulphate, and ferric salts containing organic polymers are examples of iron coagulants.

PAC has emerged as a major coagulant. The advantages of PAC include its proper dispersion, no insoluble residue, unaffected settling, more effectiveness than alum, and less space (maybe about 50%). The disadvantage of PAC is that it is less effective in removing colour.

2.4.3.1.2 Coagulant aids

Coagulant aid is an inorganic material employed in conjunction with the main coagulant. Coagulant aids, when applied, boost the density and toughness of the flocs, preventing them from breaking during the mixing and settling processes. Bentonite, calcium carbonate, sodium silicate, and polyelectrolytes are popular coagulant aids. Polyelectrolytes are polymers with absorbable groups that build bridges between particles or charged flocs. These polyelectrolytes replace the colloid's ionic group and allow hydrogen bonding between the colloid and the polymer. Polyelectrolytes are classified into three types:

(i)   cationic (adsorbed on negative colloids),
(ii) anionic (adsorbed on positive colloids), and
(iii) non-ionic.

2.4.3.2 Flocculation

The process of generating destabilised particles (or particles created due to destabilisation) to come together, make contact, and form larger agglomerates is called flocculation. After the coagulation, the flocculation process is carried out to form bigger aggregates to be settled down during sedimentation. The typical flocculation chamber in a water treatment plant is shown in Figure 2.9. When micro flocs collide, they clump together to form bigger visible flocs. The floc size grows due to additional collisions and interactions with new inorganic or organic polymers. Coagulant aids help bridge, bind, and reinforce the floc, increase weight, and speed up the settling process. Once the floc has reached its optimal size and strength, the water is ready for sedimentation. Design flocculation contact times range from 15 to 20 min to an hour or more. The amount of mixed energy and the mixing velocity must be carefully monitored. The mixing velocity and energy are generally reduced as the floc size grows to prevent the floc from ripping apart or shearing.

Figure 2.9 Flocculation chamber in water treatment plant (courtesy of Nagpur Municipal Corporation and Orange City Water Private Limited, Nagpur, India).

There are two steps to the flocculation process. The first, known as peri kinetic flocculation, is a naturally random occurrence when water is allowed to move (Brownian movement). Because Brownian motion has no effect beyond a certain floc size, flocculation occurs almost immediately after instability and is completed within seconds. Furthermore, while the Brownian movement's kinetic energy can overcome the potential energy barrier between colloidal particles, as the particles coalesce, the magnitude of the energy barrier grows roughly proportional to the area of the floc, so perikinetic flocculation of such potentially repellent particles must eventually come to an end. Orthokinetic flocculation is the second step of the flocculation process, caused by induced velocity gradients in the liquid. Setting the liquid in motion can be carried out through baffles or mechanical agitation inside a flocculation reactor or a convoluted route through interstices of a granular filter bed or differential settlement velocities inside a settling basin, which may produce such velocity gradients. The impact of velocity gradients inside a liquid body establishes relative velocities between particles, allowing interactions. The velocity gradient provided to a flocculating system is the main factor determining the rate of orthokinetic flocculation. Both applied velocity gradients and flocculation time determine the degree or extent of flocculation.

2.4.3.3 Sedimentation

Sedimentation is a water treatment process that removes suspended particles such as flocs, sand, and clay from the water. Figure 2.10 shows a typical sedimentation unit in a water treatment plant where water is kept to settle the sediments. During the sedimentation process, solid particles can settle by gravity on the bottom of settling tanks, also known as sedimentation tanks

Figure 2.10 Sedimentation tank in water treatment plant (courtesy of Nagpur Municipal Corporation and Orange City Water Private Limited, India).

or primary clarifiers. This process includes lowering the velocity of water, which causes suspended particles to settle owing to gravity. Under the action of gravity, a discrete particle in a fluid settles. The particle accelerates until the fluid's frictional drag force equals gravity, at which time the particle's vertical (settling) velocity remains constant. When turbulence is decreased by providing storage, suspended particles in water with specific gravity larger than water tend to settle down by gravity. The depth of the sedimentation tank has no bearing on the particle size that can be eliminated in the settling zone. The overflow rate is the design factor of the sedimentation tank and corresponds to the terminal settling velocity of the completely gone particle. Sedimentation tanks may work on an intermittent or continuous basis. The theoretical average duration for which the water is kept in the settling tank is the detention period or hydraulic retention time. Examples of sedimentation/settling units are horizontal flow settling tanks, tilted plate settlers, tube settlers, lamella settlers, and floc blanket installations are examples of sedimentation/settling units.

2.4.3.3.1 Types of settling

Type I: Discrete particles – Particles settle singly without interacting with surrounding particles.

Type II: Flocculent particles – Flocculation causes particles to gain bulk and settle quickly.

Type III: Hindered or zone settling – Mass of particles tends to settle as a unit, with individual particles staying in fixed locations.

Type IV: Compression – Particle concentration is so high that sedimentation can only occur by structural compaction.

2.4.4 Filtration

Filtration is important in removing naturally occurring and treatment-induced particles from water in the treatment plant. Figure 2.11 shows a typical filter house in a water treatment plant. Filtration is based on a mix of physical and chemical phenomena; the most significant is adsorption, as shown in Figure 2.12. Suspended particles directly contact the surface of individual media grains or previously deposited material when water passes through the filter bed and adsorb (stick). The forces at work in coagulation and flocculation are the same ones that attract and hold the particles to the grains. The filter bed does experience some flocculation and sedimentation. This demonstrates the significance of effective chemical coagulation prior to filtration.

2.4.4.1 Filter materials

Sand, whether fine or coarse, is often used as a filter medium. Gravel may support the sand layers, enabling filtered water to flow freely to the bottom drains and wash water to flow evenly upwards. Anthrafilt is sometimes used instead of sand as a filter media. Anthrafilt is made from anthracite, which burns with no smoke or flame. It is less costly and has a high filtering rate.

Figure 2.11 Filter house in water treatment plant (courtesy of Nagpur Municipal Corporation and Orange City Water Private Limited).

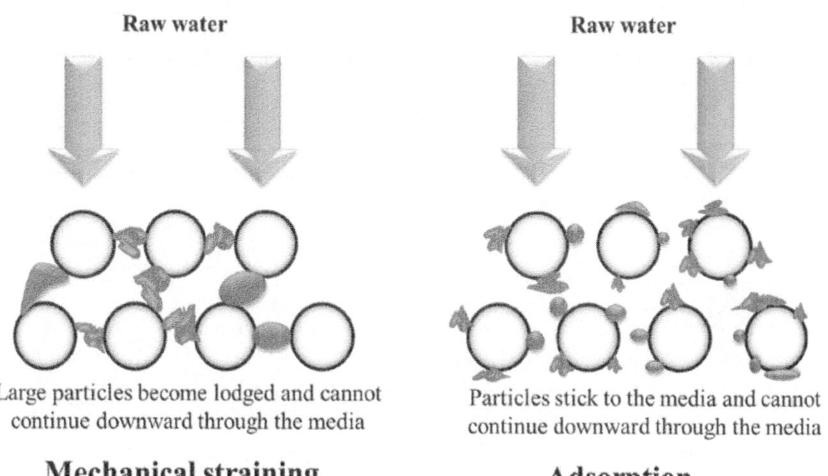

Figure 2.12 Mechanism of filtration.

2.4.4.2 Types of filters

The most typical filtering unit is a rapid sand filter (RSF), consisting of a layer of sand on top of a bed of graded gravels. Gravity filtering is done regularly to eliminate these elements. In this technique, water is routed through layers of sand and gravel. The force of gravity pushes the water through the medium. The solid particles in the water get trapped in the pores, and clear water passes through the bottom of the filter.

2.4.4.2.1 Slow sand filter

Slow sand filters are made of fine sand supported by gravel. They collect particles at the bed's surface and are cleaned by scraping away the top layer of sand containing the particles. Backwashing water through the bed fluidises the particles and cleans them. The multimedia filter comprises two or more layers of various granular materials of varying densities. Anthracite coal, sand, and gravel are often utilised. Combining the several layers may create a more flexible collection than a single sand layer. Because of the density disparities, the layers remain cleanly separated even after backwashing. Slow sand filters are being faced due to many operational limitations.

2.4.4.2.2 Rapid sand filter

RSF comprises bigger sand grains supported by gravel and captures particles distributed across the bed. The filtration mechanisms of an RSF are nearly identical to those in a slow sand filter, as shown in Figure 2.13. Moreover, the biological processes involved in other filtration are minimum in the RSF. This is due to the substantially shorter filtration runtime required between the cleanings, hindering mature biological growth. A reverse water flow is used to remove pollutants from the RSF sand bed, preceded by some agitation. The major components of the RSF include fitter media (sand bed), support gravel, underdrain system, and so on. The sand bed is made up of pure silica sand (depth between 0.60 and 0.75 m), and the used silica sands are of the effective

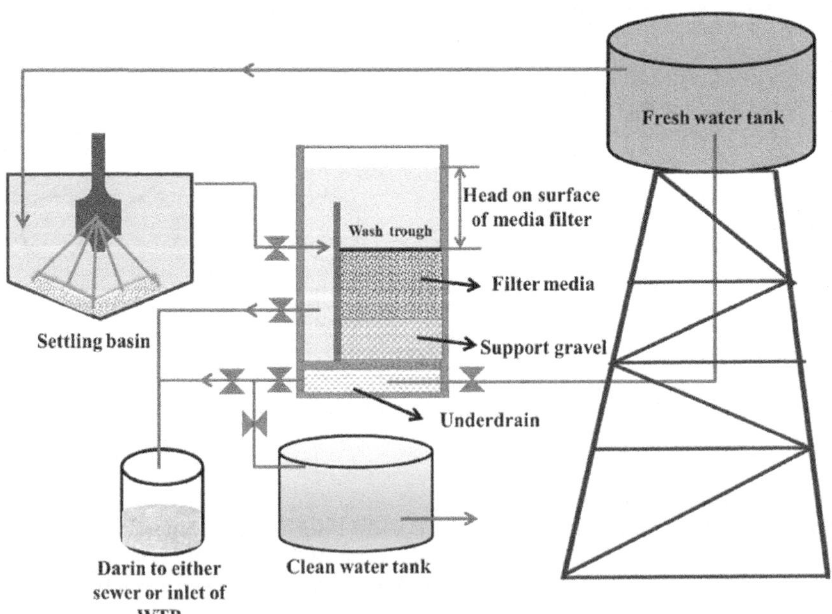

Figure 2.13 Rapid sand filtration.

size of 0.9–1.0 mm. The gravel support ensures that filtered water is consistently removed from the sand bed. The contaminated groundwater is carried via the underdrain system, which is hydraulically designed. As a result, cleaning is completely automatic and hydraulic, that is no human resources are required; hence the cost decreases. RSF works 10 times faster than slow sand filters. Moreover, the RSF required a lesser land area than slow sand filters.

Disinfection is the final step in the water treatment plant before water is transported and distributed to consumers. Chlorination is the most common disinfection process used in conventional water treatment plants. However, UV radiation and ozonation are increasing, particularly in developed countries. Details of disinfection processes are described in Section 2.5.5.

2.5 TREATMENT FOR GROUNDWATER

Groundwater generally remains less contaminated except for geo-genic contaminants such as arsenic and fluoride. While microbial contaminants remain a significant concern due to unhygienic conditions around groundwater sources, such as tubes well fitted with handpumps, chemical contaminants make it different from surface water. Fluoride and arsenic are priority contaminants in groundwater, and the treatment technologies described below are related to these two contaminants.

2.5.1 Pre-filtration (optional)
Pre-filtration, often known as screening or coarse filtration, is a popular method of removing larger particles and bringing down turbidity from membrane filtration systems. It involves the designated filter to remove particulate matter (dirt, sediment, etc.) from the water before it is treated further. There are two reasons why pre-filtration is essential. For example, the presence of larger particulate matter has the possibility to clog or abruptly exhaust the filters after it is in the system. Second, in the presence of particle matter, the efficacy of subsequent treatments (chemical or mechanical) can be greatly diminished. Pre-filtration ensures the long-term performance of membrane systems by supplying high-quality feed water.

2.5.2 Electrocoagulation
The use of electrocoagulation to treat groundwater for drinking has piqued the interest of the environmental sector. The electrocoagulation process involves directly applying current/cell potential to sacrificial metallic anodes, which dissolve the metals. Further, its precipitates as oxides and hydroxides depend on the electrolyte pH. The electrocoagulation process is primarily used to remove arsenic and fluoride from groundwater using coagulation and flocculation processes (Figure 2.14). Electrocoagulation produces chemical species *in situ*. Hence neutralisation of excess chemicals normally added in other processes is not required. In an electrocoagulation reactor, iron/aluminium/alloy electrodes are used. The sacrificial anode is dissolved when a direct current is applied, and hydrogen gas is produced at the cathode. Aqueous metallic species are generated based on anode and solution chemistry, which chemisorb arsenic or

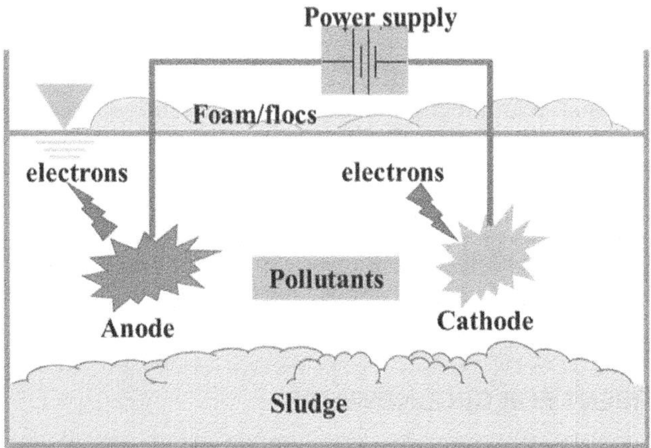

Figure 2.14 Electrocoagulation process.

fluoride in the water. These coalescing flocs became large and were removed using settling and subsequent filtration. Sludge generation in this process is low.

2.5.3 Adsorption

The adsorption method has become one of the most popular treatment procedures for reducing organic pollutants from groundwater (Figure 2.15). Adsorption has advantages like higher removal efficiency, cost-effectiveness, less energy requirement, low waste formation, and so on. It utilises adsorbents that adsorb contaminants from groundwater. The adsorbent may be natural or synthetic, and the natural adsorbents are primarily planted products or their derivatives. Even at very low concentrations, several adsorbents (activated

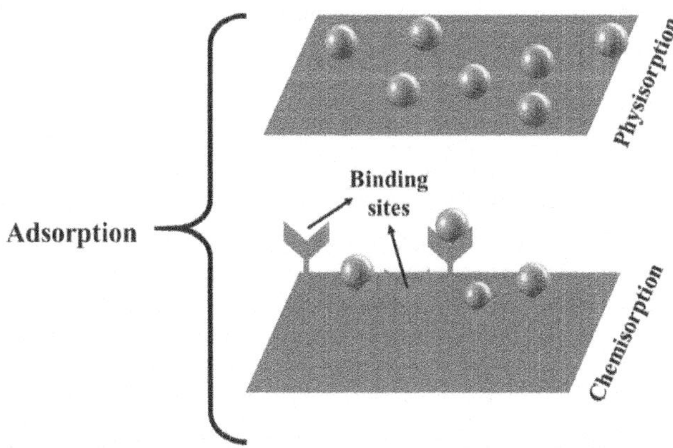

Figure 2.15 Adsorption process.

carbon, chelating resins, fly ash, and those derived from natural sources) can be employed to remove the contaminants (heavy metals, pesticides, microorganisms, etc.) from the contaminated groundwater. Activated carbon is the most commonly used adsorbent because of its affinity to adsorb the contaminants' verities and low energy utilisation, and lesser maintenance costs. However, zeolites are also utilised as adsorbents because of their homogenous pore distribution and polar bonding sites.

2.5.4 Membrane filtration

Membrane filtration removes dissolved chemicals from solutions, and tiny particles from solutions can constantly separate and concentrate colours from effluents. This approach offers unique properties compared to other methods, such as resilience to temperature rise, harsh chemical conditions, and microbial attack. Membrane filtration effectively treats effluents with low dye concentrations and recycles textile wastewater, but it is unsuccessful at reducing dissolved solid content, making water reuse problematic. Microfiltration (MF), ultrafiltration (UF), reverse osmosis (RO), and nanofiltration (NF) are the four most popular membrane separation procedures. Low energy usage, simplicity, and environmental friendliness are the key advantages of membrane technology over alternative separation techniques. Membrane filtration is more effective than traditional pretreatment since membrane pretreatment technologies use much less space and chemicals than traditional pretreatment technologies. Membrane filtration can be used in a variety of water and wastewater processes. Membrane filtration technology removes bacteria by combining the effect of physicochemical interactions between the membrane and microorganisms with the sieving effect. Membrane filtration has been discussed in detail in subsequent chapters.

2.5.5 Disinfection

Any method of destroying, inactivating or preventing the growth of microbes in water is known as disinfection. Inactivation is accomplished by modifying or eliminating critical structures or processes within the microorganism. Denaturation of proteins, nucleic acids and lipids are examples of inactivation processes. Disinfectants are the chemical agents used to inactivate or kill pathogens on an inert surface. The factors affecting the efficiency of disinfectants are shown in Figure 2.16.

2.5.5.1 Chlorination

Chlorination is adding chlorine (Cl_2) or chlorine compounds such as calcium hypochlorite ($Ca(OCl)_2$) and sodium hypochlorite (NaOCl) to water. Because of its efficacy in killing and removing disease-causing microorganisms such as viruses, bacteria, and protozoa (to a limited extent), chlorination is widely utilised as a preferred technique for disinfecting water for human consumption. The effectiveness of chlorine in killing pathogens is measured in terms of the 'CT value,' which is the product of the chlorine concentration (mg L^{-1}) and the duration of contact (min).

Figure 2.16 Factors affecting the efficiency of disinfectants.

Hypochlorous acid (HOCl) and hypochlorite ion (OCl⁻) are the most effective residuals. Chlorine hydrolyses to generate HOCl, as indicated below

$$Cl_2 + H_2O \rightarrow HOCl + HCl$$

The unstable HOCl dissociates to form hypochlorous and hypochlorite ions, as shown below:

$$HOCl \leftrightarrow H^+ + OCl^-$$

The temperature and pH of raw water are the regulating parameters in the breakdown of HOCl. As the pH rises over 7.5 (20°C), a greater proportion of free chlorine is in the form of a hypochlorite ion. HOCl and OCl⁻ are known as 'free available chlorine in the water'. Figure 2.17 shows various stages of chlorination. As complete oxidation occurs, the amount of chlorine is low in the first stage, with no free residual chlorine, reducing the final components available in the water. The synthesis of chloramines and chloro-organics occurs at this stage, demonstrating that the amount of chlorine injected is proportionate to the overall amount of chlorine residue. When the amount of chlorine is increased further in the third step, the oxidative destruction is complete, and chloramines are oxidised. As a result, the amount of residual chlorine drops to a point known as the breakpoint or dip. Because there are no further reactions due to the added chlorine after this point, residual chlorine continues to rise. The breakpoint concentration is determined by the raw water quality and the desired residue.

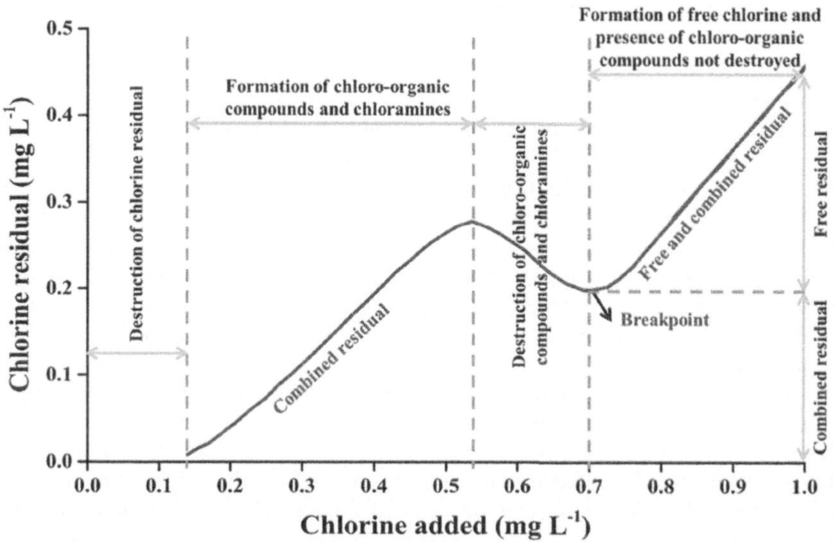

Figure 2.17 Various stages of chlorination.

The interactions between disinfection chemicals (HOCl and OCl⁻) and certain enzymes caused by microorganisms' presence in raw water distort the metabolism and eventually eliminate the microbes. HOCl is a significantly stronger disinfectant, oxidant, and reactive agent than OCl⁻ but consumed considerably higher. HOCl is neutral, penetrating negatively charged bacterial surfaces and suspended particles more readily.

If the water includes organic matter, nitrites, iron, manganese, or ammonia, chlorine instantly reacts with these compounds. These compounds are reducing agents, and no residual can occur until the chlorine reacts with these agents. Breakpoint chlorination refers to assuring the availability of free chlorine in public water supply systems. The addition of chlorine right after the breakpoint is critical because it persists as residual free chlorine. As a result, sufficient chlorine addition is required to meet the chlorine demand of water, the reaction with ammonia, and residual free chlorine for further water treatment and protection.

Chlorine is not suitable for every situation; it has several limitations. Firstly, chlorine does not kill every disease-causing microorganism in water; pathogens like *Cryptosporidium* are extremely resistant to chlorine and its derivatives. Secondly, when chlorine is added to water for disinfection, it reacts with organic matter to form disinfection by-products containing high concentrations of trihalomethanes (THMs) and haloacetic acids (HAAs).

Several scientific studies have found a direct link between the presence of disinfection by-products and adverse health effects. Some of the health effects mentioned include the risk of colon and bladder cancer, the risk of infant development reduction and unfortunate birth results in humans, the development of tumours in the kidney and liver of animals, loss of infertility and pregnancy in animals (Jolley *et al.*, 1983). However, there is no conclusive and substantial evidence for any of the aforementioned effects. According to WHO (2020), when compared to the risks associated with poor disinfection, the risks posed by these by-products are quite low. In order to avoid the hazards associated with poor disinfection, it is vital that disinfection efficacy should not be compromised.

2.5.5.2 *Ultraviolet disinfection*

Ultraviolet (UV) is known to be an effective disinfectant as a water treatment technology because of its strong germicidal (inactivating) potential; UV is powerful enough (ionising radiation) to break down chemical bonds and pathogenic organisms. Microbiological development in water can be effectively reduced by UV radiation. Bacteria, viruses, algae, and fungi can be killed by UV light at a wavelength of 254 nm because their deoxyribonucleic acid is destroyed. UV water disinfection has numerous advantages over conventional sterilisation processes, such as chlorination, for many water disinfection applications. Because deoxyribonucleic acid plays such an essential role in the activities and reproduction of microorganisms, breaking it hinders them from being active and proliferating. To ensure that all suspended particles are eliminated from the water before it is treated with UV radiation, it is necessary to filter them first.

2.5.5.3 *Ozonation*

Ozonation is a chemical water treatment in which ozone is injected into wastewater. Compared to chlorination, ozone has a higher disinfection efficacy against bacteria and viruses over the wide range of water pH. Moreover, it does not require any chemicals. Furthermore, the oxidising characteristics of the ozone decrease iron, manganese, and sulphur levels in the water and can mitigate or eliminate taste and odour issues. The iron, manganese, and sulphur in the water are oxidised by ozone to produce insoluble metal oxides or elemental sulphur, further eliminated by post-filtration. Ozonation has several applications since it is effective for disinfection and the decomposition of organic and inorganic contaminants.

2.5.5.4 *Copper–silver ionisation*

The ancient process of disinfecting water using copper and silver metals has been used in water treatments. Copper–silver ionisation is a kind of disinfection process, and this is primarily used to control *Legionella*, the bacteria responsible for the Legionnaires' disease. This ionisation technique uses positively charged copper and silver ions to generate electrostatic connections with microorganisms' negatively charged cells, thus destroying them. Apart from destroying the microorganism, the copper–silver ionisation is due to advantages like less running cost, no change in the water test, user-friendly and requiring less space.

2.5.6 Solar disinfection

Solar disinfection (SODIS) is a cost-effective, user-friendly, and environmentally friendly household water treatment system promoted as a convenient process for disinfection. SODIS refers to neutralising microorganisms in microbiologically polluted water by exposing them to sunlight (Figure 2.18). It is one of the

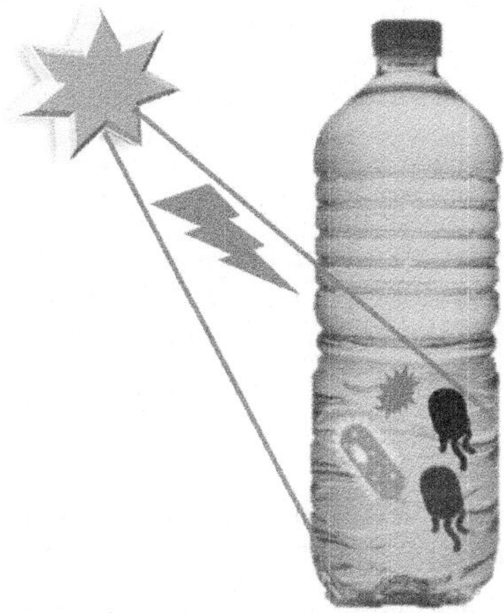

Figure 2.18 Solar disinfection.

significant alternative household water treatment technologies that rely on the germicidal effects of solar radiation and heat, in which raw water is placed into polyethene terephthalate (PET) bottles. The exposure of the PET bottle to sunlight may vary from 6 to 48 h, depending on the solar intensity and the pathogen's susceptibility. It has been found that the SODIS method destroys waterborne pathogens. Sunlight-exposed PET bottles can be reused for over 6 months, and no undesirable by-products, such as plasticisers, are detected (Mintz et al., 2001).

2.6 DEGRADATION OF WATER QUALITY AT THE SOURCE

Diffuse-source pollutants and the geographical and temporal variability associated with these sources are the main causes of water quality decline. Water treatment processes are complicated when source water quality deteriorates, resulting in higher treatment costs and lower tap water quality. Moreover, it has recently come to light that drinking water sources may also include pharmaceutical products. These, too, are substances created to influence living beings. These medications are excreted through urine and faeces after usage, ultimately getting into municipal wastewater. Microplastics are another type of contaminant. Because this is often particle stuff, additional treatment is necessary to deal with it. They cannot be degraded or absorbed by conventional methods, and the particles are frequently too small (micro-and nanoscale) to be removed by conventional filtration procedures such as sand filtration.

According to their source and passage to the receiving environment, water quality degradation at the source is typically classified as point or diffuse. This distinction is critical in terms of water quality legislation and pollution control.

Point sources of pollution, such as pipes and ditches from sewage treatment plants, industrial sites, and confined intensive livestock operations, release pollution directly into receiving water bodies at a discrete place. Point source pollution has the greatest influence on water quality during the summer and dry seasons when river flows are low, dilution capacity is low, and during storm seasons, when combined sewer overflows are more common.

During seasons of rainfall and irrigation, **diffuse sources** of pollution are indirectly discharged to receiving water bodies via overland and subsurface flow, atmospheric deposition to surface waters, and leaching through the soil structure to groundwater. When rainfall promotes hillslope hydrological processes and runoff of pollutants from the land surface, the most severe water quality consequences from diffuse source pollution occur during storm periods (especially after a dry spell).

2.7 DETERIORATION OF WATER QUALITY FROM SOURCE AND CENTRALISED WATER TREATMENT PLANTS TO HOUSEHOLDS

Although many efforts are being undertaken to identify safe water sources, maintaining these sources' water quality sustainably is a major challenge. While catchment management is a daunting task to protect surface water sources, overexploitation and poor aquifer protection adversely affect groundwater sources for community water supply. Wright *et al.* (2004) reported inferior water quality at the PoU compared to the source in half of the studies considered in their review article. In the same review article, none of the studies reported improved microbiological water quality at the PoU. Shield *et al.* (2015) undertook a detailed analysis of variation in water quality in terms of faecal contamination at the source and household stored water and determined the relationship between contamination at each sampling point and type of water supply. They considered 319 articles that reported water quality data at the source and household storage water. It is reported that water quality significantly deteriorates between source and stored water. The mean percentage of contaminated samples at the source was 46% as against 75% samples in household storage water. Piped water supplies had significantly lower contamination than non-piped water, primarily due to residual chlorine.

Elala *et al.* (2011) investigated the common risks through the water supply chain in Nagpur, India and reported a significant reduction in water quality from source to use as shown in Figure 2.19. Bain *et al.* (2014) undertook a meta-analysis to compare faecal contamination of water from 'improved' sources with that from 'unimproved' sources and assessed variation in contamination by source types. Water quality between countries varied substantially as <0.01% of samples from piped supplies in Jordan were contaminated with thermotolerant coliforms (TTC) compared to 9–23% of piped supply in the rest of the countries in which rapid assessment of drinking water quality was studied. The proportion of water samples in piped supply having *E. coli* was substantially higher in rural (61%, $n = 101$) than urban (37%, $n = 1470$) settings in Peru. Water from 'improved sources' is less likely to contain microorganisms of faecal origin than unimproved sources. However, improved sources cannot be considered consistently safe. Although groundwater sources were comparatively safe, high contamination levels were occasionally observed in boreholes.

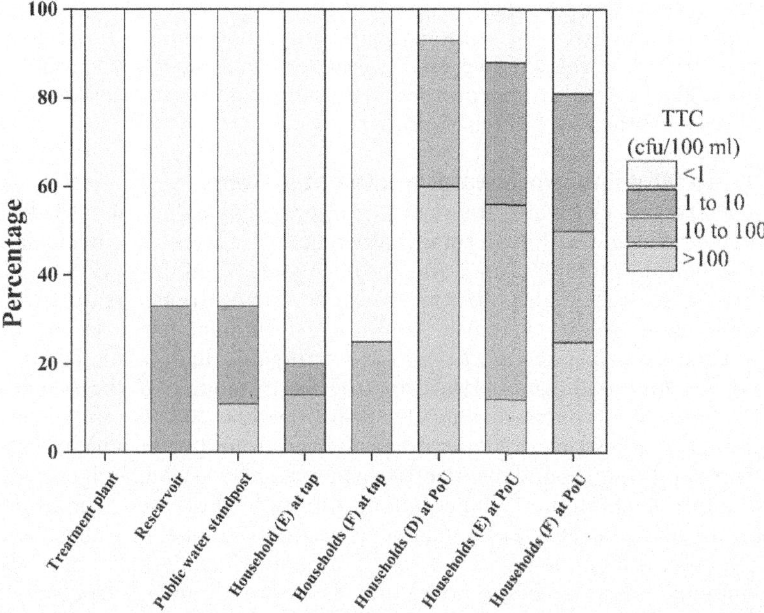

Figure 2.19 Thermotolerant coliform count down the supply chain in the water distribution network. (Elala et al., 2011).

Another study in Nagpur, India, indicated gradual degradation in microbial water quality from the treatment plant to the consumers while surface water sources also had poor water quality. It was also reported that water quality at the household storage deteriorates most in the low-income groups, as 80.6% of the samples were found contaminated. Survey findings among consumers indicated that the water at PoU was contaminated due to unsafe management. Households having PoU treatment units were less prone to contamination (John *et al.*, 2014).

Seifert-Dähnn *et al.* (2017) also assessed water quality (faecal coliforms) changes from source to consumption in 135 households in Maharashtra, India. Although 98% of the households received water from the improved source (surface water, open wells, and boreholes), about 50% of sources and even more during the monsoon had faecal coliforms (from 'not detected' to 6500 CFU/100 mL). Surface water sources and open wells had maximum faecal contamination, whereas boreholes had few faecal coliforms. About 51% of households consumed water having faecal contamination. Significant temporal and local variation in water quality at the water source and the point of consumption was reported. A study of 1726 water sources and 1676 stored water samples in Bangladesh indicated that 41% of sources were positive for *E. coli*. 89% of household water samples had *E. coli* in control (without any safe storage and chlorination) as compared to 70% (safe storage) and 26% (both safe storage and chlorination) (Ercumen *et al.*, 2015).

Source water of higher quality is more likely to get contaminated, whereas water quality remains poor without treatment in households. In households having intermittent water supply, quality is adversely affected at the initial stage of water collection. It is important to ensure safe water at sources as this may affect hygienic practices in households.

2.7.1 Need for point-of-use water treatment systems

The primary goals of water treatment are to prevent pathogen-caused acute infections, reduce the long-term negative health effects of chemicals and microbiological contaminants, and provide safe drinking water that is conditioned so that quality does not degrade during transmission from the treatment plant to the customer. Traditional treatment processes primarily aim to remove suspended particles, natural organic matter, dissolved iron, manganese, and pathogens. However, the performance of these methods for the removal of nitrates, fluoride, heavy metals, PPCPs, and emerging contaminants is restricted. General aeration of groundwater will necessitate filtration, conditioning, and disinfection, whereas surface water will necessitate coagulation/flocculation, sedimentation, filtration processes, conditioning, and disinfection. Because polar chemicals are more soluble in water, they are typically more difficult to remove during treatment. Flocculation's effectiveness in removing organic micro-pollutants is often limited, especially for pharmaceuticals and endocrine-disrupting chemicals and insecticides and their metabolites (endosulphan, ethylene thiourea, and 1,2,4-triazole). Slow sand filters are effective at removing germs but not at removing viruses. Chemical disinfection can kill viruses, but it is less effective in water with many ammonia and nitrogen because these chemicals react with hypochlorite and quench it. To overcome this limitation, a higher amount of chlorine is added, which increases the development of hazardous disinfection by-products and changes the taste of the product water, making it unpalatable to customers. Because microbial pathogens are shielded by suspended particles in turbid water, UV disinfection and SODIS are ineffective because UV radiation is insufficient to inactivate pathogens under non-ideal sunlight intensity conditions. In addition, none of these approaches addresses chemical contamination directly.

Hence, it can be concluded that water received in households often remains contaminated due to poor source water quality, inefficient conventional water treatment, compromised water distribution network and inadequate plumbing systems. Membrane-based systems require less specialised input for operation than older treatment methods, while qualified technicians are required for repairs. If properly handled and maintained, treatment should be able to eliminate viruses and some chemical contaminants from a wider range of source water conditions. Unlike single-barrier techniques for water treatment, multibarrier membrane-based methods offer built-in redundancy that reduces the risk of failure. Adopting an onsite water treatment and storage system is a viable alternative in these places. PoU technologies are onsite water treatment systems that eliminate pathogens in water sources before consumption. Several new technologies have emerged in the last few years and are being used in PoU water treatment systems.

2.7.2 Summary

A water supply system comprises catchment, sources, treatment plants, distribution networks, and consumers. Source remains the major concern from a water quality perspective. In the absence of a water safety plan, the conventional water treatment plant is the only protection for consumers. Conventional water treatment consists of processes, viz., aeration, coagulation, flocculation, sedimentation, filtration, and disinfection, although retrofitting and addition of units such as UF are added to such plants. Groundwater remains less contaminated except for geo-genic contaminants to avoid microbial contamination if maintained properly. With the degradation of water quality of the sources, conventional water treatment plants often fail to meet drinking water quality standards with respect to many contaminants.

A very high proportion of households, particularly in developing countries, have contaminated water supply despite having safe water sources and centralised water treatment. Contamination while transporting, collecting, storing and handling in households in many instances is observed to be negating the effects of safe water source identification and treatment. There exist many reasons for water contamination in households despite the safe water source, adequate conventional treatment and non-leaking distribution network. This also has provided an opportunity to discuss comparative advantages of water treatment at source against the PoU. Moreover, degradation of source water quality with contaminants that cannot be removed in centralised conventional water treatment plants requires additional treatment to meet stringent drinking water quality standards. With sequential deterioration in water quality from centralised conventional water treatment plants to households, treatment at PoU appears inevitable.

doi: 10.2166/9781789062724_0059

Chapter 3

Point-of-use water treatment systems

3.1 INTRODUCTION

Point-of-entry (PoE) or PoU water treatment systems overcome possible contamination while transporting and distributing water from the centralised treatment plant. Besides treating the water before it enters the reticulation system at the consumer sites, PoE water treatment systems are placed. As a result, each family will require one PoE water treatment system. PoE water treatment systems are less expensive than centralised treatment plants, especially when reticulation expenses for rural societies are factored in. PoE water treatment systems can have a processed water storage tank to regulate maximum flow rates, or they can be designed to withstand peak flows without it. Because they are installed on the property, the treated water does not remain in the reticulation system for longer durations, and the possibility of further contamination is limited. The fundamental issue with PoE water treatment systems is that constant monitoring of water conditions is not financially viable. Rather than processing all incoming water to a property, PoU water treatment systems treat the water at a single tap. These systems handle a relatively small quantity of water and are typically installed beneath the kitchen sink, on the wall or at the platform/table-top providing only that tap with treated water. PoU water treatment systems are available with various treatment technologies, including filtration and disinfection.

PoU water treatment systems can be classified into two categories (1) treatment during storage in a container which is mainly applicable when the household is not connected to piped water supply; (2) treatment in households having access to piped water supply and internal plumbing also facilitates plumb-in system. PoU water treatment systems do not require a piped water supply, and electricity is generally based on treatment processes such as filtration and chemical addition (e.g. addition of coagulants/disinfectants) (Pagsuyoin *et al.*, 2015). On the contrary, systems such as RO membrane-based water treatment systems require a piped water supply and electricity to operate the pump to generate the requisite pressure to drive the treatment system. With increasing awareness about water safety, many technologies or their variants (including multiple technologies often termed as multi-barrier approaches) are merging. There are other driving factors for the increasing growth of PoU water treatment systems.

PoU water treatment system is used in households preferably prior to consumption and is mainly intended to remove pathogens and chemicals such as arsenic in some instances. Several technological solutions have been developed in the last few decades and implemented as PoU water treatment systems. There is a rational change in the use of PoU water treatment systems in developing countries, and people who can afford them are demanding systems that can remove almost all the contaminants. Considering such requirements and improved awareness about water contamination and adverse health effects, demand for PoU water treatment systems, particularly those based on membrane processes, is increasingly being offered by manufacturers and vendors. In some developing countries, membrane-based PoU water treatment systems are now considered synonymous with PoU water treatment systems.

3.2 POINT-OF-ENTRY AND POINT-OF-USE WATER TREATMENT SYSTEMS

Water purification is crucial to make it potable and fit for usage and consumption. Water from ground wells and other water bodies such as rivers and lakes is collected in reservoirs. Water purification systems are sometimes installed on the property before the water enters the home. These are called PoE water treatment systems. These systems can consist of single or multiple technologies to address issues related to the incoming water. PoE systems are much larger than PoU systems because they handle higher flow rates and volumes.

3.2.1 Point-of-Entry water treatment systems

PoE water treatment systems are installed either outside immediately before the water enters a household or in an interior location inside before the water is distributed throughout the household. PoE water treatment systems may comprise large carbon filters, water softening/conditioning systems, or other technologies to remove chemicals, and odour, neutralise aggressive water or eliminate suspended solids and sediments. These systems make the water fit for general usage, and further treatment may be required to make the water suitable for drinking or cooking.

These PoE water treatment systems (also called pre-filtration or pre-filters) considered level I treatment systems are used before the water is again treated at a location where further specific treatment is required. If a higher level of treatment is not required throughout the household, it can significantly reduce the cost by using smaller systems downstream before the water is used.

The PoE water treatment systems are usually large and durable and can last several years without major servicing. These are cost-effective systems; however, routine cleaning or maintenance may be required to increase their operational life.

3.2.2 Point-of-use water treatment systems

This type of water treatment system is intended to be used at a specific household location. Many can be installed under a kitchen sink or directly attached to a tap or faucet. However, plumb-in and the wall systems are increasingly being installed. More advanced systems use a storage tank under the sink with filters installed on an adjacent wall. These systems treat the water before drinking and cooking. At this stage, typically, water is treated at least for the second time as it is already treated at the conventional water treatment plant. However, in some instances, particularly in developing countries, water is directly used from underground/surface sources, and PoU remains the only treatment option. These are essential systems in households and commercial places. These systems improve water quality to make it suitable for drinking and cooking. In this case, the water consumed is treated just before its usage.

The fact is that both systems come with their benefits, and one is not better than the other. Whether one should utilise PoE or PoU depends on requirement and budget. The basic difference here is that a PoU water treatment system is applicable for only one tap or faucet in most cases. Hence, if someone is looking to treat only drinking water, an RO membrane-based PoU water treatment system is more beneficial. Utilising the same technology for the entire home would be very expensive, and one may not need highly purified water for laundry or cleaning dishes.

In the case of a PoE system, which is perhaps installed at the main supply line, all the taps and faucets in the household would receive filtered, purified water. This may be useful if the supply line source is well water containing heavy metals and smelly chemical compounds. It is a known fact that hard water leaves residues, stains, and discolouration on utensils, clothes, bathtubs, and showers. In such a scenario, all the water sources in the home should receive filtered water.

Ideally, these systems are essential in most cases and need to be installed at their levels. This chapter focuses on the PoU water treatment system.

3.3 CLASSIFICATION OF POINT-OF-USE WATER TREATMENT SYSTEMS

PoU water treatment systems are divided into several categories, as depicted in Figure 3.1. A single tap or a limited number of taps is connected to the PoU water treatment systems for consumption, mainly for drinking and cooking purposes. PoU water treatment systems are popular for low- and moderate-income communities because of their ease of use, low cost, less maintenance, and independence from the local water and power grid. Using the principle of size exclusion, filtration removes microorganisms having a size larger than the filter's pore size. RO is the most used technology in membrane-based technologies because of its wide range of capabilities to remove contaminants, including microorganisms. UV disinfection is a widely accepted and validated technology suitable for various industries, including electronics and municipal wastewater disinfection. While treating water for drinking purposes, sometimes we need to add some vital minerals. With the remineralisation of RO water, we will get those (vital) micro-nutrients without contaminating the water supply

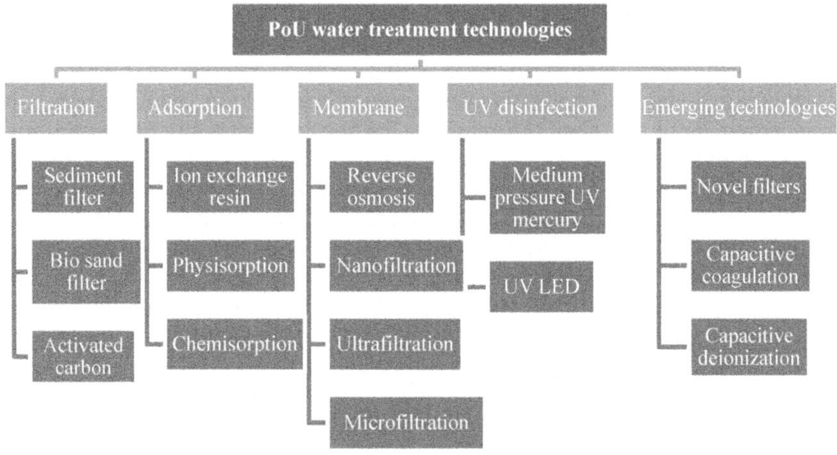

Figure 3.1 Classification of PoU water treatment systems.

(these micro-nutrients, vital minerals, and dissolved solids can be supplemented through food).

Based on the position of the PoU water treatment system, it can be classified into different categories as discussed below:

- **Plumbed in PoU system:** The plumbed-in form of the PoU system necessitates a permanent connection to the household's existing water pipes, forcing filtered water through the connected sink faucet. A separate tap is piped in using a comparable system and fitted in the same manner. The difference is that the water in the second system is distributed through a dedicated auxiliary faucet, which is frequently located near the kitchen sink.
- **Countertop PoU system:** Tubing connects the countertop PoU system to a kitchen faucet. The treated water is then distributed from a separate spout in the system or the kitchen faucet. A countertop manual fill device is set up on the counter and activated when someone pours water into it. This system is capable of treating water in small amounts.
- **Faucet PoU system:** On an existing kitchen faucet, a faucet mount system is fitted. When drinking water is required, a diverter directs water through the system. Unfiltered water is delivered through the ordinary kitchen sink faucet when not in use.
- **Pour-through PoU system:** Gravity drips water through a filter and into a pitcher, generally kept in the refrigerator, in pour-through water treatment systems. Pour-through systems can normally only filter a small amount of water at a time due to their tiny capacity. This type of system must be refilled regularly.

Table 3.1 lists the main treatment mechanisms and contaminants that these technologies can address and an assessment of their relative effectiveness and the most likely treatment location (PoU, source).

Table 3.1 Relative effectiveness of water treatment technologies for treating select pollutants.

Technology	Pathogens	Turbidity	Heavy Metals
Boiling	★★★		
Cloth filtration	★	★★★	
Natural coagulant/flocculants	★★	★★	★
Biochar			★
Ceramic water filters	★★★	★★	
Sand filtration	★★★	★	★★★
Chemical disinfection	★★★		
UV irradiation	★★★		
SODIS	★★★		
RO filtration	★★★		★★★

3.4 COMPONENTS OF MEMBRANE-BASED POINT-OF-USE WATER TREATMENT SYSTEMS

Even though the membrane process is straightforward, the overall system is frequently complex. Figure 3.2 depicts a typical PoU water treatment system, including pre-treatment filtration, membrane, flow regulator, post-treatment unit, and storage tank. Pre-treatment helps the system last longer by eliminating suspended solids and chemical compounds such as chlorine, which can otherwise damage the membrane. Post-treatment can eliminate any taste- and odour-causing compounds or residual organics that the membrane process did not remove. Another purpose of post-treatment, particularly in RO membrane-based systems, is mineralisation/addition of dissolved solids to treated water. Multiple ways can achieve remineralisation component to make up for the lost dissolved solids. Monitoring gauges, sensors and lights are becoming more popular in PoU water treatment systems. Shut-off valves are essential for stopping

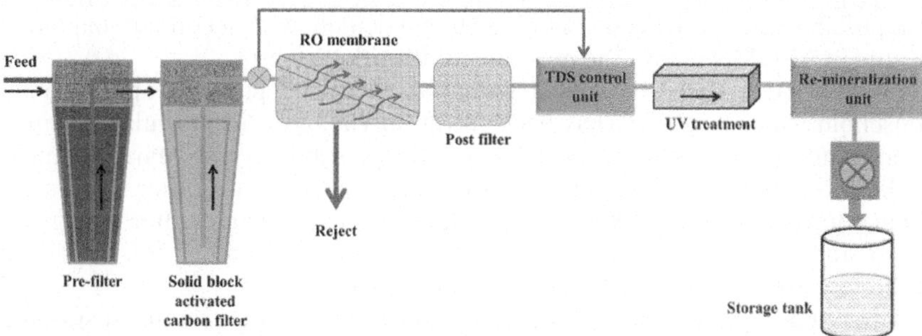

Figure 3.2 Components of PoU water treatment system.

water flow when the storage tank is full, preventing squandering surplus water. Multi-barrier water treatment systems such as disinfection through UV, and remineralisation, specifically in membrane processes, are common in PoU water treatment systems. Because membrane processes consume a comparatively more quantity of water, the adequacy of the residential sewage system should also be considered, although reuse of reject water from RO-based system is also being attempted. Many variants of PoU water treatment systems are available primarily due to marketing strategy rather than any other necessity.

3.4.1 Storage tank

The storage tank has a capacity of 8–20 L in most cases. When the tap is open, it is pressured to ensure proper flow. This tank is usually provided to store water to provide a buffer in case the membrane-based PoU water treatment system is not temporarily in operation. The storage tank is also useful wherever only intermittent water and power supply are available, which is the case in most developing countries. Water can be treated during the availability of water and power (if the system is power-driven), and treated water can be stored in the tank. This flexibility has improved the acceptability and provision of membrane-based PoU water treatment systems.

3.4.2 Pre-treatment

The deposition of colloidal particles, organic matter, or suspended solids on the membranes leads to membrane fouling, causing a dense cake layer formation. The formation of a dense cake layer decreases membrane flux. This requires frequent cleaning of membranes in PoU water treatment systems, and sometimes termination of the operation and membrane replacement becomes inevitable. Pre-treatment processes such as pre-filtration, and so on, are necessary to avoid this. Most household pre-treatment filtration utilises the mechanical filtration system. In the mechanical filtration system, mostly larger suspended particles, including sand, silt, clay, loose scale, and organic matter, are removed from the water. Removing dissolved chemicals or tiny particles with these filtration techniques is impossible. However, while combined with other treatment systems, the mechanical filtration system can also remove dissolved chemicals or fine particles. The removal of iron, manganese, and hydrogen sulphide is also possible after converting dissolved to particulate with secondary oxidation treatment installed before the mechanical filtration system.

Since ancient times, the most extensively utilised physical approach for household water treatment has been filtration via a porous granular medium, often sand or successive anthracite coal and sand layers. Many granular media filters are available, being utilised for household and other small-scale applications. This filter includes bucket filters, drum or barrel filters, roughing filters, and so on. Sand, charcoal, crushed sandstone or other soft rock, and anthracite are granular media used in water filtration.

Three types of mechanical filtration are utilised in water treatment systems, that is, cartridge sediment, single media and multimedia filters. The quantity and particle size of the suspended solids in the contaminated water and the rates at which water must be processed decide the type of mechanical filtration appropriate for the household. A sand media filter, for example, can remove contaminants faster than other types of filtration systems, but it is not appropriate for removing tiny particles.

The pre-treatment of the contaminated water is essential to achieve mainly the following objectives:

- Eliminate carryover of suspended solids or colloidal particles, or organic matter into membrane pores.
- Modify the characteristics of the organic matter in the water being treated. Their deposition potential is also reduced on the membrane surface and within pores.
- Removal of chlorine from feed (raw) water.

Following are the three major classes of pre-treatment filters used in PoU water treatment systems.

3.4.2.1 Media filters

These are pressure filters and can be filled with the user's household preferred filter media. In these filters, both sand and carbon media have been used. An autonomous backwash valve has been included in the units, controlled by the data acquisition and control unit. Backwash can alternatively be controlled by a timer or a pressure drop over the filter, which is a less expensive option while using PoE water treatment systems. There are several manufacturers of such devices, and the efficiency is determined by the filtration rate, filter media depth, and type. Backwash valves that operate automatically are also widely available.

3.4.2.2 Cartridge filters

Cartridges filters are available in various pore diameters (0.5–50 μm) and filter lengths. Absolute pore size ratings are available on these filters (cloth and paper filters). Because the paper or cloth has a somewhat constant pore size, similar to a membrane filter, separation of particles larger than the given absolute pore size can be guaranteed. Solids and particles are separated as they pass through a thick open filter, and particles are separated as they collide with the filter material. Cartridge filters without an absolute pore size rating work similarly to a depth filter. When particles pass through a thick open filter and collide with the filter medium, thus they are removed. A wrapped string or foam material can be used to create these filters.

3.4.2.3 Activated carbon filters

Activated carbon (AC) and its derivatives are mainly used for pre-treatment, particularly in membrane-based PoU water treatment systems. AC has a very high surface area of $1000\ m^2\ g^{-1}$. Various raw materials are used to produce a variety of AC with varying characteristics. AC filters are divided into the following two categories:

(i) Granular activated carbon (GAC).
(ii) Carbon block filters.

The application of AC filters is the preferred option due to its versatility and, more importantly, its capability to remove residual chlorine from feedwater. As a membrane-based PoU system is increasingly used in the settings having water supply through a centralised treatment plant, the presence of chlorine is possible in the feedwater. Hence, chlorine needs to be removed before feedwater is received by the membrane unit, which can be achieved through AC.

If other media filters are used in membrane-based PoU water treatment systems, the AC filter should be integral to the pre-treatment (filtration) system. Hence, the design of the pre-treatment unit is limited to the AC filter in Chapter 4.

3.4.3 Membranes

As water contamination increased in the last few decades, membrane processes emerged as a possible solution in water treatment due to flexibility in removing almost all the constituents from water. A membrane is a thin, semi-permeable sheet that separates constituents of water depending on their size and phase (liquid/gas). Membrane technology for the PoU water treatment system has gained much attention because of its excellent manufacturing quality and capability to remove almost all the constituents from water. Because of the safety and ease with which they can be used, membrane processes are deemed cost-effective and practical. Compared to technologies such as flocculation/chlorination, membrane filtration systems are simpler to operate.

Membrane processes are available in various configurations and separate constituents from water. Most separation processes used in water treatment systems are being replaced by one or a combination of membrane processes. The process efficiency is usually measured in terms of two factors: (1) how effectively the membrane separates the constituents and (2) how fast the water can be processed (treated) per unit area of the membrane. In an ideal scenario, highly efficient separation with a high rate of removal of compounds should be achieved, but, in reality, one objective is generally compromised to attain the required performance with regard to the other. Membrane science and technology have a significant history of research and development in the laboratory before making their first substantial industrial use in the 1960s. Membrane-based processes now have many applications and significantly improved human lives after nearly 50 years of advancement (Figure 3.3).

The membrane remains the core and essential component of membrane-based PoU water treatment systems, as other components may be found in other PoU water treatment systems. The driving force in the membrane process is the pressure differential between the two sides of the membrane. Water and the small-molecular-weight solutes with smaller pore diameters pass across the pressure-driven membranes under the influence of static pressure. In contrast, the large molecular weight solutes are not allowed to pass across the membranes of the high-pressure to low-pressure side of MF, UF and NF membranes (Figure 3.4). In the case of the RO membrane, only water passes through, while the membrane on the high-pressure side retains almost all types of constituents (solutes).

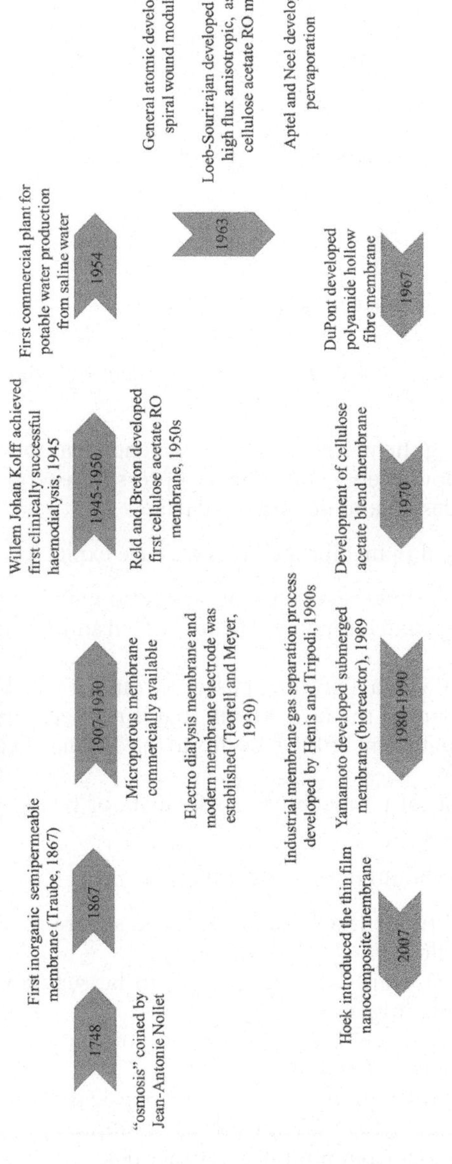

Figure 3.3 Historical development of membrane technology (*sources: Fane et al., 2011; Krishnan et al., 2022*).

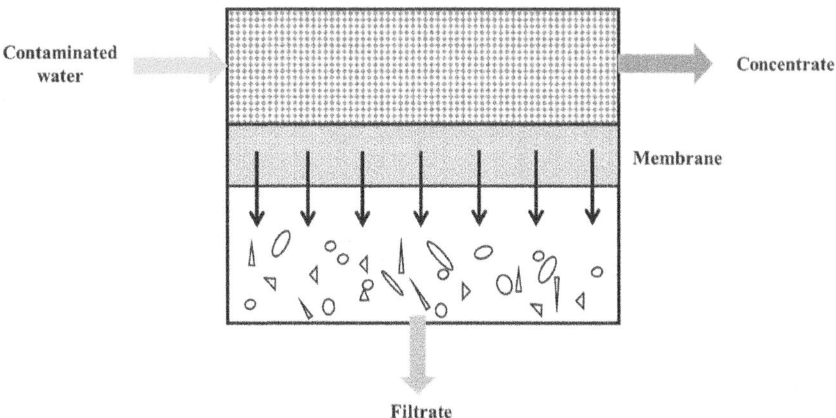

Figure 3.4 Principle of membrane process.

Membranes have dominated water treatment technologies due to multiple advantages:

- Allow controlling the permeation rate of a chemical species carefully.
- Are inherently low-energy consumers (if pressure-energy is available).
- Have a simple design and no moving parts.

The basic terms used in membrane processes are explained below:

- **Feed:** stream entering the membrane separation module.
- **Permeate:** stream that is separated from the feed and crosses the membrane barrier.
- **Flux:** the rate at which a product passes through a membrane.
- **Reject:** the amount of the feed stream that is not separated.
- **Permeability:** coefficient linking the flux to the driving force (and membrane thickness).
- **Selectivity:** ratio of permeability coefficients of two species (greater or equal to 1).

A good membrane should possess the following properties:

- High permeability for the component to be separated (smaller area for given permeate flow).
- High selectivity towards the component to be separated in relation to other components (higher purity).
- The low effective thickness of the active portion of the membrane (to ensure a high permeation and low cost).
- Good mechanical strength to support the physical structure.
- High membrane stability in real working conditions.
- Uniformity-freedom from pinholes or other defects.

The basic classification criteria of the membranes are transport mechanism, nature, structure, and geometry (Figure 3.5).

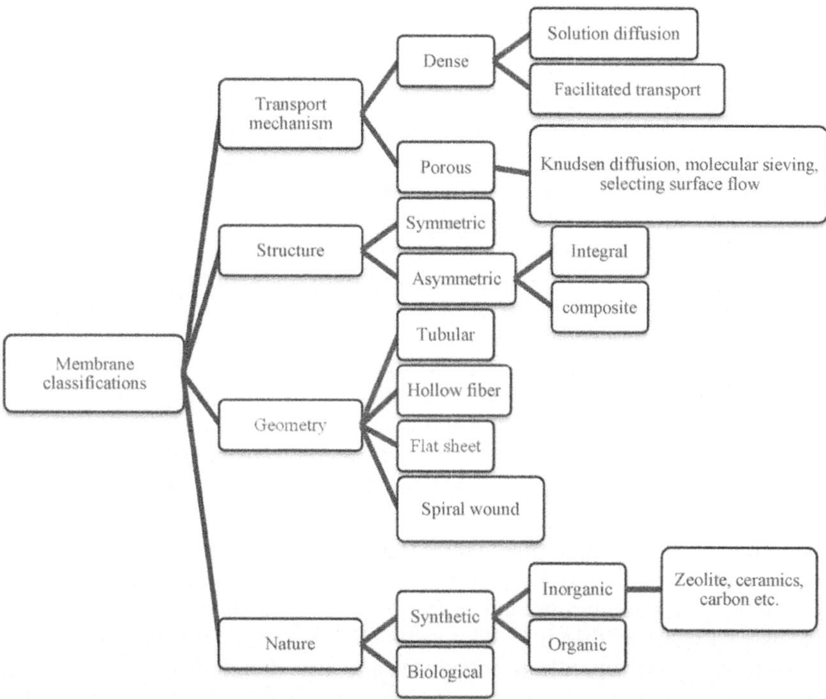

Figure 3.5 Membrane classifications.

3.4.3.1 Transport mechanism

The transport mechanism plays an important role in the membranes-based process. The membranes may be porous or dense, based on the transport mechanism involved. These membranes (porous or dense) are utilised in the different membrane filtration processes as per targeted contaminants. Porous membranes have a mean pore size diameter of 1–5000 nm, and the driving force is the pressure gradient. A non-porous membrane has a mean pore size diameter of less than 1 nm driving force is a concentration gradient across the membrane. Porous membranes are transported by viscous flow or diffusion, and the selectivity of constituents is based on size exclusion. At the same time, transport is based on the solution-diffusion mechanism in non-porous membranes.

The transport mechanism of solute (constituents in water) or solvent (water) molecules across the membranes are categorised as follows:

- Bulk flow across pores of membranes.
- Diffusion across membranes through pores (of the membranes).
- Restricted diffusion across membranes through pores (of the membranes).
- Solution or solvent diffusion across (dense) membranes.

Different classes of membranes have different transport mechanisms of solutes and solutions across the membranes. Transport mechanisms in microporous membranes, for example MF and UF, are driven by the pore size of the membranes. Hence, the aforementioned transport mechanisms (a) and (b) are observed in MF and UF membranes. An RO membrane is characteristically a dense membrane due to smaller pore sizes, and the movement of constituents across the membrane is a solution-diffusion transport mechanism. NF membranes display behaviour between porous and dense membranes. Additionally, size-exclusion and electrostatic interactions are the basic phenomena governing the constituent (solute) rejection by RO/NF membranes.

In addition, there are charged membranes in which separation takes place either with porous (fixed charged chemical groups on the pore wall) membranes or non-porous (swollen gel). This separation occurs due to charge exclusion (based on the Donnan effect in which ions/or molecules having the same charge in solute as the fixed ions in the membrane are rejected, whereas those with opposite charges are transported across the membrane). In another category of the membrane, that is a carrier-mediated membrane, transport is based on ions/molecules having a special affinity for specific substances in the feed.

3.4.3.2 Structure

Membranes are also classified based on structures with respect to cross-sections, which can be either symmetric or asymmetric (Figure 3.6). Based on structure, the membrane may be symmetric, or membranes can be divided into two categories: thin-film composite and phase-separation membranes. Anisotropic phase-separation membranes were sometimes referred to as Loeb–Sourirajan membranes after the researchers first developed them. The chemical composition of these phase-separated membranes is homogeneous, but the structure is not. Phase-inversion techniques, such as those mentioned above, are used to create Loeb–Sourirajan membranes, except that the pore sizes and porosity change across the membrane thickness. The asymmetric membranes are fabricated with the integral asymmetry, or there may be composites asymmetry. Isotropic membranes have the same composition and

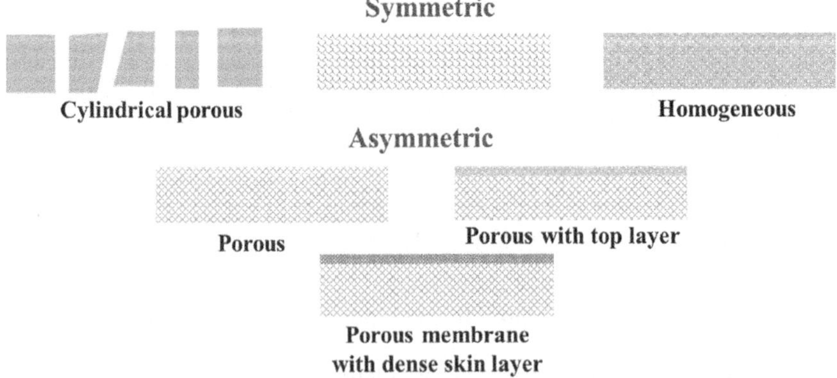

Figure 3.6 Membrane classification based on structure.

physical properties across their entire cross-section. Anisotropic membranes have layers that differ in structure and/or chemical composition and are non-uniform across the membrane cross-section.

Symmetric (isotropic) membranes can be self-supported non-porous membranes (mainly ion-exchange) and macroporous MF membranes, which can be the following membranes:

- Dense, non-porous symmetric membranes and not commonly used as the flux (treated water) is too low for all the practical purposes.
- Microporous membranes are widely used as MF membranes.

An asymmetric (anisotropic) membrane is a thin porous or non-porous selective membrane supported mechanically by a relatively thicker porous substructure. Following are the characteristics of these membranes:

- Reduces effective thickness of the selective barrier.
- Permeates flux that can be increased without change in selectivity.

An asymmetric membrane combines the high selectivity of a dense membrane and the large permeation rate of a very thin membrane. The breakthrough and development of asymmetric membranes (RO-based seawater desalination by Loeb and Sourirajan in 1963) was a critical discovery in membrane technology.

The composite membranes are made up of two or more components. In the most basic scenario, a selective membrane material is placed on a porous support layer as a thin layer. This support layer offers mechanical strength; the thin top layer is responsible for the separation. A multi-layer composite membrane comprises a porous substrate and multiple layers of various materials, each serving a particular purpose. Different types of coatings such as dip, coating and spin, and polymerisation (interfacial, in situ, plasma, grafting, etc.) are applied. A wide range of polymers can be applied to make composite membranes due to the variety of preparation processes available for thin-film NF and RO composite membranes. Composite membranes have the following key advantages over integrally skinned asymmetric membranes:

- Individual material selection for the separation layer and porous supports.
- Individual processing of the separation layer and porous sub-membrane, enabling for optimisation of each structural element.
- Because a small quantity of polymer is needed to produce the thin separation layer, highly expensive membrane materials can be used.

3.4.3.3 Geometry

The membranes are also classified based on geometry, and they may be tubular, hollow fibre, flat sheets or spiral wound (Figure 3.7).

Tubular membranes have the following characteristics:

- not self-supporting
- turbulent flow
- good resistance to fouling

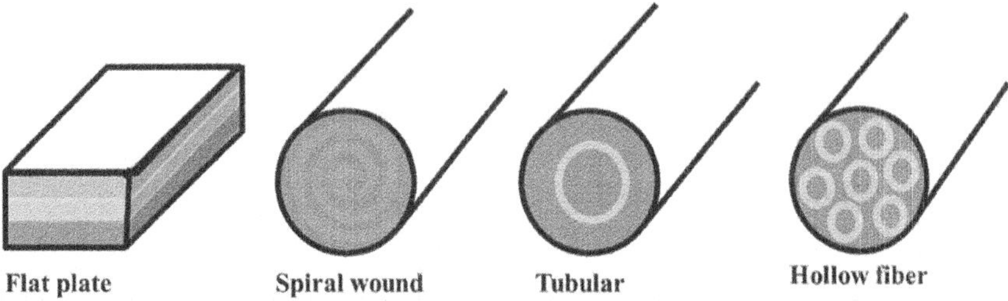

Figure 3.7 Membrane classification based on geometry.

- application limited to UF (wherever fouling and concentration polarisation are major concerns)
- high cost (low surface/volume ratio)

Hollow fibre membranes have the following characteristics:

- High surface-to-volume ratio.
- Shell-side feed for high-pressure applications (e.g. gas separation).
- Bore-side feed to avoid concentration polarisation (no stagnant volumes).

Flat sheet membranes have the following characteristics:

- Easy to clean and replace.
- Low packing density (the ratio of membrane area to packing volume).
- Better mechanical stability.

In spiral wound systems, flat sheet membranes are wrapped around a centre core in a spiral form, akin to a roll of fabric. The semi-permeable membrane is gapped with a spacer material and sealed at the edges to allow the filtered liquid to flow through.

3.4.3.4 *Nature*

Based on nature, the membranes may be synthetic or biological. Synthetic membranes are further classified as inorganic and organic membranes. The layer deposition approach can customise the generated substrate's membrane selectivity and other membrane properties.

Organic membranes are also known as polymer membranes and are normally made of cellulose, PTFE, PVDF, PP. Organic membranes have the following characteristics:

- Rigid in the glassy state or flexible in rubbery form.
- Cost-effective, good selectivity and easy processability.
- Prone to fouling and chemically non-resistant.
- Limited operating temperature and pressure.
- Short life span.

Following inorganic membranes are commonly used in water treatment applications:

- Ceramic membranes (metal oxide, metal carbide, zeolite).
- Metallic membranes (palladium and palladium alloys).
- Carbon membranes (graphene, carbon nanotubes, coal).

Inorganic membranes have the following characteristics:

- Chemically and thermally stable, mechanically strong, and capable of operating in extreme feed circumstances.
- Good temperature (up to 500°C) and water resistance, well-defined and stable pore structure, high chemical stability, long life duration.
- Fragile, rigid.

This book provides details of organic membranes as they are mostly used in membrane-based PoU water treatment systems, and details of inorganic membranes are not further added.

3.4.3.5 Pore size

The membranes are generally classified based on pore size, and the membrane industry extensively uses this classification. The size of the membrane pores determines the degree of selectivity of the membranes. They are classified as MF, UF, NF, and RO membranes based on pore size (Strathmann *et al.*, 2011). Figures 3.8 and 3.9 show the details of pore sizes, operation pressure and the contamination rejections from the different classes of the membranes.

3.4.3.5.1 MF membranes

MF membranes are the second-oldest commercially used membrane application after dialysis. MF is a pressure-driven separation method employed to concentrate, decontaminate, and separate macromolecules, colloidal particles, and suspended particles from a solution (Kuiper *et al.*, 1998). The nominal pore diameters of MF membranes are usually in the range of 0.1–1.0 mm. MF is

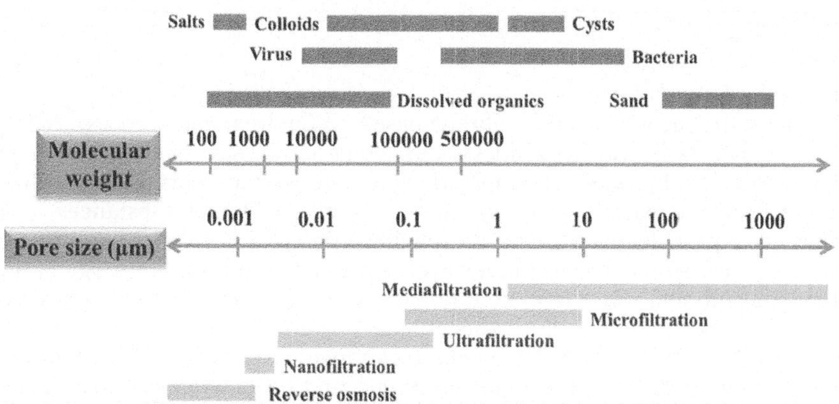

Figure 3.8 Pore size and molecular weight of membranes.

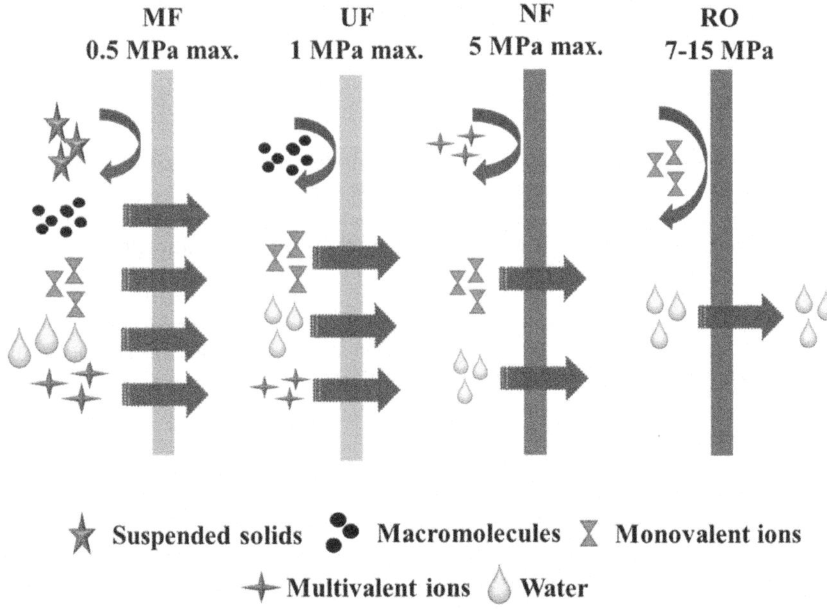

Figure 3.9 Operating pressure requirements for membranes.

typically operated at low transmembrane pressures (4 bar or 0.4 MPa). At first, cellulose-based MF membranes were used.

Now, a wide variety of membranes made of polymers and inorganic fillers are available, depending on the application. Many innovations have been suggested to improve membrane selectivity and/or reduce fouling and its detrimental consequences.

Due to the increasing application of UF, primarily because of the capability to remove microorganisms even as small as viruses, the use of MF is declining.

3.4.3.5.2 UF membranes

The most frequently used membrane-based water treatment process is UF. The UF has low land occupancy rates, consistent water quality, and high automation. Furthermore, it can nearly eliminate microorganisms from water, significantly increasing water biosafety. Viruses, colloidal substances, and suspended particles can all be thoroughly filtered. UF technology offers a greater processing efficiency, better treatment effect, and lower energy usage than conventional water treatment processes. Particulate and microbiological pollutants are removed with this technique; however, ions and tiny molecules are not. The service life of UF membranes ranges from three to five years or longer. Tubular, hollow-fibre, plate and frame, and spiral-wrapped UF modules are commercially available. Solutes of a diameter of 0.03 μm and greater are rejected by UF membranes.

There are more than 10 types of UF membrane available globally, including polystyrene (PS), polyacrylonitrile (PAN), polypropylene (PP), polyethylene (PE), and polyvinylidene fluoride (PVDF). The UF process has a greater removal rate of turbidity and particulate matter than the traditional process; effluent turbidity is constant below 0.1 nephelometric turbidity unit (NTU), and particulate matter removal rate is up to 99.9%. It can eliminate pathogenic microorganisms efficiently. In reality, UF membranes may reach reductions of 7 log total coliform bacteria, 4.4–7 log *Cryptosporidium*, 4.7–7 log *Giardia lambia*, and 6 logs or more of certain viruses such as *MS2 bacteriophage*.

3.4.3.5.3 NF membranes

NF evolved from RO and UF and was initially referred to as open RO, loose RO, or tight UF. NF is a pressure-driven membrane process that sits between UF and RO to reject the molecular or ionic species. In NF membranes, larger free space, microscopic pores, or nanovoids may be accessible for transport. The size of these nanovoids, which can range between 0.5 and 1 nm, causes a transition between microporous and dense membranes. The nominal cut-off for NF membranes is commonly assumed to be 1000 to 200 Da. Compared to UF and RO, NF has always been difficult to define and explain. NF membranes have very strong rejections for multivalent ions (99%), low-to-moderate rejections for monovalent ions (about 70%), and high rejections (90%) for organic molecules with molecular weights greater than the membrane's molecular weight.

Because most NF membranes are hydrophilic, they can filter both charged and uncharged (primarily organic) particles from water. The separation is mostly performed through a sieving action, in which the solute's molecular size must be higher than the NF membrane's pore size.

3.4.3.5.4 RO membranes

Osmosis is the natural process of fluid (water) diffusion through a semi-permeable membrane from a solution with a low solute concentration to a solution with a greater solute concentration until both sides of the membrane maintain an equilibrium of fluid concentration. The fluid's osmotic pressure is equal to the pressure difference between the two sides of the membrane. The potential chemical gradient across the membrane is the primary water flow cause. Abbe Nolet, a French physicist, devised RO in 1748. That year, he witnessed solvent flowing from a lower concentration solution to a higher concentration via a semi-permeable barrier. As a result, the scientific community became aware of the notion of osmosis. When on the membrane, the applied pressure is more than the osmotic pressure than the pure water passed through the membrane pores on the permeate side and while the high solute concentration on the other side of the membrane, resulting in the separation of water from the contaminated water. The process is known as 'RO' since it is the inverse of normal osmosis (Figure 3.10).

The primary difference between osmosis and RO is that osmosis is a naturally occurring phenomenon in which water molecules pass along a concentration gradient. In contrast, RO is a water purification process in which water molecules

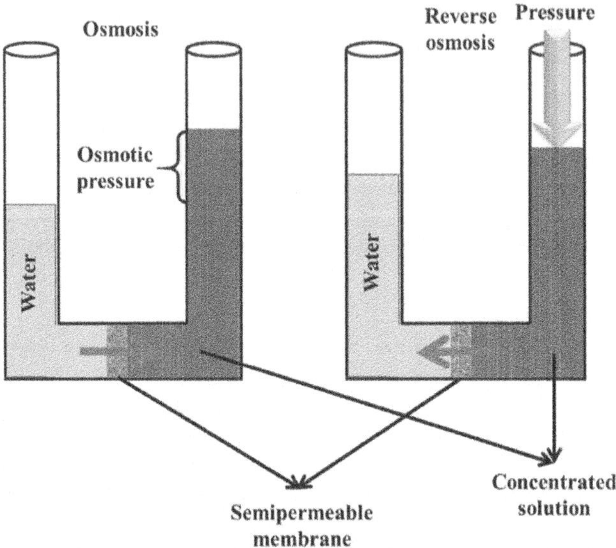

Figure 3.10 Osmosis and RO processes.

pass across a semi-permeable membrane against a concentration gradient. The process of RO is used in desalination and purifying water. It is osmosis in the other direction, as the name implies. Water is pushed through a semi-permeable membrane against the concentration gradient by applying a pressure larger than the natural osmotic pressure. As a result, water molecules move from a low water potential to a higher potential through an RO membrane. Dissolved salts, organics, microorganisms, and pyrogens, for example, will not flow through the barrier. As a result, RO makes it easier to filter water in water purification procedures. Unlike osmosis, RO requires an energy input to apply pressure to water.

Although MF and UF membranes are often made of the same polymers, they use distinct processes, resulting in differing pore sizes. MF and UF polymers include PVDF, polysulfone (PSF), poly (ether sulfone) and PAN copolymers. MF membranes consist of cellulose triacetate-cellulose nitrate mixtures, nylons, and polymers. RO membranes are typically made of cellulose acetate or PSF coated with aromatic polyamides. NF membranes, like RO membranes, are made of cellulose acetate blends or polyamide composites, but they can also be modified like UF membranes by using sulphonated PSF.

3.4.4 Post-treatment

Regulatory restrictions, system design, projected water quality parameters, and water chemistry influence the chosen post-treatment technique. Stabilisation, disinfection, and corrosion control are common post-treatment operations, although they might also involve degasification and/or air stripping. Blending, remineralisation, disinfection, and the materials utilised for storage are the four main concerns with post-treatment water. To increase their acceptability and, in particular, to minimise their aggressive attack on materials, desalinated waters are frequently blended with tiny amounts of more mineral-rich fluids (WHO, 2004). The water used for blending should be completely safe to drink.

Membrane filtrations are a pressure-driven process that uses membranes as a physical barrier to keep large-molecular-weight compounds while permitting water and low-molecular-weight substances (less than their cut-off) to pass through. Consequently, the quality of the water following post-treatment is strongly dependent on the membrane-based processes used.

After the proper treatment, some groundwater or surface water can be used for blending, enhancing the hardness and ion balance. Blending various and disparate water supplies, with desalted water as one of the suppliers, is becoming more common. The amount of TDS and hardness in water affects its taste. Water with a low TDS level is flat and tasteless, whereas water with a high TDS level (>2000 mg L^{-1}) becomes disagreeable and unappealing. Palatability is regarded as good, with TDS levels up to 600 mg L^{-1}.

Post-treatment is required depending on where the process water is used at the PoU; for example, RO permeate is remineralised prior to distribution for potable use. It is critical to treat the permeate once it has been collected to avoid corrosion of downstream pipelines and equipment. The RO product water is slightly acidic (pH 5–6), soft, and low in alkalinity (Brandt *et al.*, 2016), whereas distilled water is near-zero hardness and has alkalinity, unlikely to surpass 2 mg L^{-1}. These waters corrode metal and asbestos cement pipes and absorb calcium from mortar-lined pipes.

3.4.4.1 Ultraviolet disinfection

As a multi-barrier approach, a UV light to kill pathogens in the water is used post-treatment. The application of UV disinfection becomes essential if a portion of raw water is added to treated water to compensate for the loss of minerals. UV disinfection in such cases can only be used for the portion of raw water being blended into treated water instead of disinfecting the entire treated water. The primary advantage of UV treatment is that it disinfects water without using chemicals.

UV treatment inactivates (kills) microbes rather than removing them from the water. When UV light enters a microbe, its energy damages the cellular activity of the microbe, preventing it from growing. All viruses, bacteria, and protozoa are generally killed by UV radiation. Some microbes, such as *Cryptosporidium* and *Giardia lamblia*, have strong or impenetrable cell walls that low-power UV light systems cannot penetrate (Okpara *et al.*, 2011). Exposure time, lamp intensity, and general water quality characteristics all affect the efficiency of this technique. Unlike chemical disinfectants such as chlorine, UV does not influence the water's flavour, odour, or chemical characteristics, resulting in safe, palatable water.

Other disinfection treatment technologies such as chlorination and ozonation are rarely used in membrane-based PoU water treatment systems.

3.4.4.2 Remineralisation

Remineralisation allows for the recovery of certain micronutrients while also guaranteeing that the water is contaminant-free. Calcium, magnesium, and trace minerals are the most common micronutrients found in mineral water. Magnesium is a key co-factor and activator of over 300 enzymatic activities, whereas calcium is a significant component of bones and teeth. Sodium and potassium aid in the contraction of muscles and the proper functioning of nerves. Physical function suffers as a result of insufficient amounts of these chemicals. The water treated by membranes usually tastes 'bland' and unpleasant. On the other hand, alkaline water contains re-added minerals, which improve the taste.

Mineral water also successfully quenches thirst. Electrolyte-fortified water helps quench thirst faster than plain water. Because membrane-treated water has a neutral pH level, it tastes flat and dull. Because it is the perfect balance of quality and filtration, remineralised water tastes better. People are also more likely to drink water that tastes better, increasing water consumption and keeping consumers well hydrated.

Remineralisation imparts unique properties to water. The chemical properties of raw water (or any other soft water) must be adjusted, notably in buffering capacity, total hardness component concentration (Ca^{2+} and Mg^{2+}), and corrosion-related factors. The water must be post-treated in order to attain such benefits. A typical PoU water treatment system consists of at least three filtration stages – an AC filter, the membrane filter, and a post-treatment filter. Some systems with multiple filtering stages add a post-AC filter or a remineralisation water filter. When water goes through the post-mineral filter, minerals are added back to the water to normalise the pH level.

Two types of systems neutralise corrosive water:

- An automatic backwashing filter calcite (calcium carbonate) media requires annual calcite additions and complete calcite replacement every two or three years.
- Using a feed pump system to inject sodium carbonate (soda ash) into the water requires a mixing solution to fill the pump two or three times a year.

A frequent remineralisation procedure is to mix desalinated water with source water or partially treated water. The purity of the water sources used for mixing is crucial in microbiology and mineral content. The volume of water used for mixing can vary from less than 1 to 10%. Another way of remineralisation is to directly add some essential ions to desalinated (treated) water to get the desired balanced mineral concentration. Among the chemicals used are carbon dioxide (CO_2), lime ($Ca(OH)_2$), sodium bicarbonate ($NaHCO_3$), sodium carbonate (Na_2CO_3), and calcium chloride ($CaCl_2$). Because of its limited solubility and high cost, sodium bicarbonate is rarely used. Moreover, sodium bicarbonate tends to cake at high moisture levels and is difficult to store in humid conditions.

3.4.4.3 *Total dissolved solids adjustment/controller*
The microorganisms and other pollutants smaller than the semi-permeable membrane's pore size are removed. Unfortunately, it also traps minerals and salts, which are useful. As a result, the treated water from membrane-based PoU

water treatment systems is devoid of the minerals found in raw water. Drinking demineralised water over an extended period may negatively impact the intestinal mucous membrane, metabolism, and mineral homeostasis, resulting in a range of bodily systems being disrupted. If a vigorous physical exertion is followed by drinking many litres of low-mineral water, severe acute harm, such as hyponatremic shock or disorientation, might result. 'Water intoxication' is the term for this ailment. While RO water is devoid of harmful contaminants, it lacks some necessary minerals for optimal health. Even a small amount of minerals ingested through drinking water could be critical. However, contrary evidence of literally no adverse impacts on health due to drinking water devoid of dissolved solids is also reported.

TDS adjustment can be carried out by diverting a portion of inlet water without RO membrane treatment and blending it with treated water to make up for the loss of these minerals/solids. However, this needs to be carefully carried out to avoid the build-up of chemical contaminants in finally treated water, as UV disinfection is only practised in raw water (being used for blending).

3.4.4.4 Sensors

Sensors can track the consistency with which PoU devices are used and provide information on how to improve it. To remind the user when to replace a filter, most 'smart' PoU water treatment systems just contain a timer, flow counter, or 'number of times used' counter. These are based on pre-determined assumptions about filter usage and water quality rather than directly measuring the quality or quantity of water. Many parameters must be evaluated using sensors to monitor water quality in a smart water filter. A variety of sensors have been created to evaluate the physical properties of water or the presence of chemical contaminants (Wu *et al.*, 2021).

3.4.4.4.1 Electrical conductivity sensor

Electrical conductivity (EC) sensors are a surrogate for TDS, oxidation-reduction potential, pH, turbidity, ion-selective electrodes, and developing optical, fluorescent, and spectrophotometric devices – are among them. By incorporating these sensors into water filter systems, consumers will be able to monitor their water quality without having to worry about silent product expiration. Furthermore, the widespread use of sensors may encourage manufacturers to increase removal efficiency and address a wider range of contaminants. Furthermore, even if sensors similar to those described below were installed in PoU water treatment systems, they would not be truly 'smart' until they communicate directly with the system owner via a smart device such as an Internet-connected phone, tablet, or PC.

3.4.4.4.2 TDS sensors

An EC sensor, which uses merely a pair of electrical contacts to measure the resistance of water by applying a modest current, is one of the most common sensors. Because dissolved ions in water allow electricity to flow more readily between the contacts, EC and TDS are intimately connected.

Conductivity sensors cannot detect uncharged pollutants such as soluble hydrocarbons, disinfection bi-products, some medicines and pesticides because they do not modify the ability of water to conduct electrical current.

3.4.4.4.3 pH and oxidation–reduction potential sensors

pH and oxidation–reduction potential sensors are two more types of water quality sensors that do not appear to be used in any commercial PoU systems. There are several varieties of pH sensors, the most popular of which is the combination pH sensor. Two electrodes take measurements on either side of a specially developed glass membrane containing a reference solution. The pH of the test solution is proportional to the measured electrical potential.

3.5 LIMITATIONS OF POINT-OF-USE WATER TREATMENT SYSTEMS

Like any other technology for water treatment, membrane-based PoU water treatment technologies also have limitations. Attempts are being made to overcome these limitations (Harris, 2005). Membrane being the significant component of such PoU water treatment systems, limitations of membranes are also applicable to such systems. Following are the major limitations of membrane systems:

- Compared to processes, membrane processes are difficult to perform in stages, for example distillation. Membrane processes often contain single-stage; however, two or three-stage membrane processes are also feasible for PoU water treatment systems.
- Pre-treatment is essential for membrane processes, and any operational issues in pre-treatment adversely affect membrane processes' performance. Turbidity (>10 NTU) normally encountered in surface water sources and chlorine, for example, can spoil membranes used in PoU water treatment systems.
- Membrane processes are energy-intensive, although gravity-based MF plants are operational. However, multi-barrier PoU water treatment systems and particularly those based on RO membranes, require energy for operations.
- Membrane-based PoU water treatment systems are comparatively expensive. Pre-treatment and post-treatment also add to the cost of such systems. Sensors which have now become a component of PoU water treatment systems, also add to the cost.
- Periodic maintenance is essential in membrane-based PoU water treatment systems.
- Membrane fouling limits the performance of water treatment systems, and fouling can be rapid due to poor water quality.
- Trained human resources are required in the maintenance of membrane-based water treatment systems.
- Membrane processes are usually operated at normal ambient temperature, and very high temperature may affect the performance of membrane-based water treatment system.
- Membrane processes are also used for disinfection. However, there is no residual disinfection possible after these processes.

- RO membrane-based PoU water treatment systems have reject water to the tune of 40–60%, which is considered as wastewater stream and discharged into household drains. This is a major limitation of such systems, although attempts are being made to improve water recovery.
- RO membrane-based PoU water treatment systems make treated water corrosive, and dissolved solids are very low. Remineralisation is being practised, which needs careful monitoring of dissolved solids in treated water.

3.6 SMART AND FUTURISTIC MEMBRANE-BASED POINT-OF-USE WATER TREATMENT SYSTEMS

Because of the rapid advancement of internet technology, many household appliance manufacturers are now offering 'smart' goods, such as 'smart' PoU water purification systems. Smart home appliances are defined in a variety of ways. The general assumption is that a smart product can be managed remotely by the user via a smartphone, tablet, or other devices. WiFi or Bluetooth® technology connects to consumers and communicates with them via an 'app' (Wu *et al.*, 2021). Smart PoU water treatment systems come in various shapes and sizes, with varying degrees of sensor integration, but information on the filter media and sensors used in domestic water treatment systems has been scarce. Furthermore, different manufacturers appear to hold opposing viewpoints on the usefulness of water filters. Some products claim to be 'smart' because they can offer better-quality water, but they do not meet the connectivity requirements of other smart home equipment.

The internet of things (IoT)-enabled water treatment systems gradually to establish a new technology trend in the water treatment domain. These IoT-enabled water treatment systems, including PoU systems, are designed to improve and replace conventional membrane-based and other water treatment systems. With the possibility of a substantial reduction in the costs of sensors, other hardware units and support software, these smart PoU systems will be efficient with improved operation and maintenance. Moreover, these IoT-enabled PoU systems will be compact, convenient and compatible with mobile phones. These systems will be able to dispense the requisite quantity of water remotely with mobile phones. In addition, the application of sensors helps communicate with the user about the possible replacement of treatment units such as a membrane. The prohibitive cost of these systems remains a major bottleneck in spurring growth in demand, but this will surely be in use in most households employing PoU water treatment systems.

Emerging PoU water treatment technologies, such as capacitive coagulation, capacitive deionisation, and a novel class of filters, recently gained much attention.

3.7 SUMMARY

PoU water treatment system is used in households preferably to remove pathogens and chemicals before consumption. Several technological solutions are available and included in PoU water treatment systems. PoU water treatment systems are being used for several advantages, for example, low cost, less maintenance cost, and, more importantly, independence from the local water and power grid. RO is the most preferred technology among membrane-based technologies because of its versatility in removing almost all the contaminants, including microorganisms. UV disinfection is a widely accepted and validated technology provided water is devoid of chemical contaminants. Filtration is also used in PoU water treatment systems. Various configurations of membrane-based PoU water treatment systems are available. This typically includes a pre-treatment filtration, membrane, a flow regulator, a post-treatment filter, a storage tank, and a dispensing faucet.

AC and its derivatives are mainly used for pre-treatment, particularly in membrane-based PoU water treatment systems. Membranes are classified based on the transport mechanism, structure, geometry, and nature. The membranes are generally classified based on pore size, and the membrane industry extensively uses this classification. The size of the membrane pores determines the degree of selectivity of the membranes. They are classified as MF, UF, NF, and RO membranes based on pore size. Post-treatment is required depending on where the process water is used at the PoU; for example, RO permeate is remineralised prior to distribution for potable use. It is critical to treat the permeate once it has been collected to avoid corrosion of downstream pipelines and equipment.

Innovation occurs in membrane-based PoU water treatment systems, and new tools such as IoT enable the system to become smart. Concerns about such a large proportion of rejected water from the RO membrane-based water treatment system attract innovators. Other derivatives of membrane processes are also emerging, making these systems even more efficient and affordable.

doi: 10.2166/9781789062724_0083

Chapter 4
Design of membrane-based point-of-use water treatment systems

4.1 INTRODUCTION

With the advancement of membrane technology, the design and engineering of membrane-based PoU water treatment systems are also refined. Membrane configuration should be investigated first to comprehend system design, as module design is always adapted to the membrane properties. There are considerable differences in membrane chemistry (cellulose triacetate – CTA and thin-film composite – TFC with polyamide barrier layer) and, more crucially, membrane topologies (fine hollow fibre and flat sheet).

Recent advancements and continuous innovation in spiral wound module design are helping to bring down the cost of RO technology, making it more widely available for water treatment in many regions worldwide. Hence, RO membrane has become the preferred choice even in membrane-based PoU water treatment systems. This chapter describes select engineering features of the spiral wound module, pre-treatment, post-treatment, and membranes, as designed to treat water.

The spiral-wound membrane-based PoU water treatment system is a flat-sheet, cross-flow device despite its cylindrical form. The water is transported axially through the system, while the permeate travels in a spiral, radial path to the permeate collection tube. The membrane that separates these streams remains the system's technological core, but other parts of the system engineering are equally important.

The increased focus on membrane-based PoU water treatment system engineering is partly motivated by a desire to save cost and, more commonly, to get the most out of the latest membrane technology. By focusing on energy efficiency and preventing membrane salt rejection, membrane benefits can be

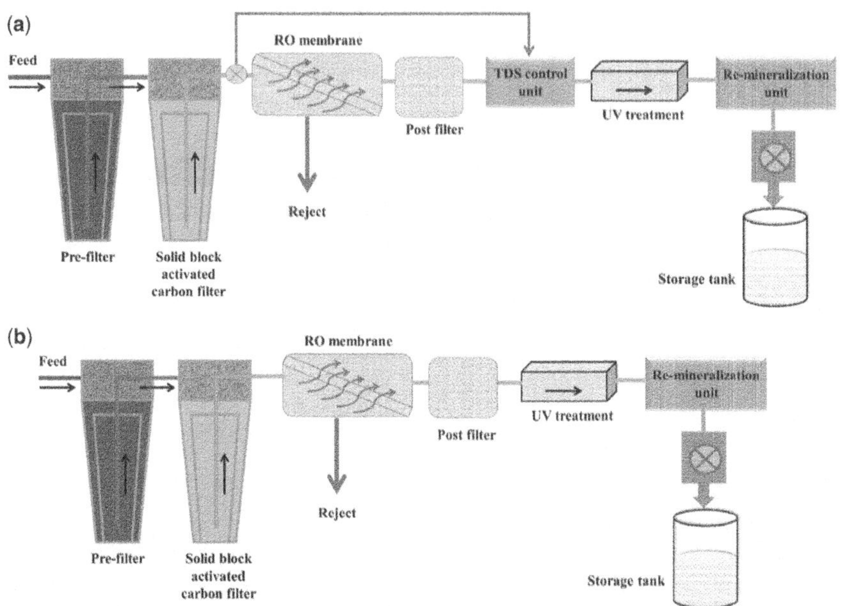

Figure 4.1 Components of membrane-based PoU water treatment system **(a)** with and **(b)** without TDS controller unit.

fully realised. Following are the major components of a membrane-based PoU water treatment system (Figure 4.1) considered for the design of the system (these components are slightly different from those presented in other chapters);

- Pre-treatment (AC filter and so on)
- Membrane
- Membrane module
- Post-treatment (UV, chlorination)
- TDS adjustment/controller (bypass, the addition of chemicals etc.)

The design of a membrane-based PoU water treatment system to successfully provide safe drinking water for a 10–20-year period should consider many aspects. Substantial data are available to design large membrane-based systems (e.g., desalination plants). However, design criteria for membrane-based PoU water treatment systems are restricted. At times, design criteria for desalination plants are modified and used in designing membrane-bed PoU water treatment systems. An attempt is made in this chapter to design a membrane-based PoU water treatment system by referring to the data available for large systems and by considering other relevant data (e.g. water consumption). The following steps are used for designing a membrane-based PoU water treatment system.

4.2 DESIGN PARAMETERS FOR MEMBRANE-BASED POINT-OF-USE WATER TREATMENT SYSTEM

The subsequent section deals with important design parameters for membrane-based PoU water treatment systems. To better understand the design of the membrane-based PoU water treatment system, we have taken two examples (hypothetical) from each design criteria. The water processing sector in the United States of America utilises some assumptions to standardise capacity calculations (Patel *et al.*, 2020). Hence, assumptions have been made while designing the PoU water treatment system.

- **Daily drinking water needs:** Each person consumes 5 L of water per day for drinking purposes.
- **Daily cooking water needs:** Five persons in the household use 10 L of water per day for cooking.

4.2.1 Capacity

Estimation of daily water demand ('capacity') is required when choosing a membrane-based PoU water treatment system. Knowing capacity or flux (volume of output water per hour) requirements can aid individuals and families in selecting an appropriate membrane-based PoU water treatment system with sufficient output capacity. Several assumptions can be used to calculate the capacity requirements of the PoU water treatment system. Various criteria are available to estimate individuals' water requirements, which vary from country to country. In addition, WHO has also specified water requirements for hydration. Individuals' water consumption varies depending on climatic parameters such as ambient temperature. These PoU water treatment systems are increasingly used in schools, hospitals, and commercial establishments such as restaurants, shops, and so on. Hence, PoU water treatment systems of various capacities are being manufactured to cater to different consumers, and the capacity of these units is mainly driven more by market forces than engineering estimation.

The following example demonstrates the application of water consumption data in arriving capacity of the PoU water treatment system.

The total volume of water can be estimated by using the following equation:

$$\text{Total requirement per day} = [\text{Number of persons in household}$$
$$\times 5\,\text{L of drinking water per person per day}] \qquad (4.1)$$
$$+ 10\,\text{L cooking water}$$

4.2.1.1 Case study 1

If there are five persons in a household, then the per day total water requirement can be estimated as follows:

$$\text{Total requirement per day} = [5 \times 5\,\text{L of drinking water per person per day}]$$
$$+ 10\,\text{L cooking water Total volume per day} \qquad (4.2)$$
$$= 35\,\text{L of water}$$

4.2.1.2 *Case study 2*

If there are 10 persons in a household, then the per day total water requirement can be estimated as follows:

$$\text{Total requirement per day} = [10 \times 5 \, \text{L of drinking water per person per day}] + 20 \, \text{L cooking water} \tag{4.3}$$

$$\text{Total requirement per day} = 70 \, \text{L of water} \tag{4.4}$$

The capacity of the PoU water treatment system is expressed in L h^{-1}, and the system is also designed on the quantity of output water per hour. As water consumption varies hourly and maximum consumption might occur during food preparation, particularly during morning hours, the design assumes that the total quantity of water will be consumed in 6 h.

Hence, the capacity of the PoU water treatment system can be considered as $70/6 = 11.67$ L h^{-1}. For practical purposes, the capacity of the PoU system can be considered as 15 L h^{-1}. This example demonstrates that the design principle used for the water treatment plant may be applied to the capacity determination of the PoU water treatment system. As there is a large variation in water consumption, PoU water treatment systems of various capacities are available to cater to different consumers.

4.2.2 Pre-treatment system

Pre-treatment of the feedwater entering an existing membrane-based PoU water treatment system can minimise fouling of the membrane and thereby increase the system's overall recovery rate. Depending on its source, the feedwater may contain suspended and dissolved solids, including organic and inorganic compounds. Suspended solids can settle/adsorb on the membrane surface, obstructing feed channels and increasing system friction losses. Scaling can occur when dissolved solids precipitate out of the water. Water treatment entering the membrane component of the water treatment system might impair the pump's efficiency, resulting in higher energy consumption.

Sediment filters are used for pre-treatment; however, these filters are gradually phased out due to numerous advantages of AC media. AC and its derivatives are mainly used for pre-treatment. Most AC is derived from natural materials such as nutshells, wood, coal, and petroleum. AC typically has a surface area of 1000 m^2 g^{-1}. However, various raw materials produce a variety of AC with varying hardness, specific gravity, pore and particle sizes, surface areas, ash, and pH (Rivera-Utrilla *et al.*, 2011). Because of these differences in properties, certain carbons are preferred over others in various applications. AC filters (pre-treatment units) can be divided into

- GAC and
- Carbon block filters.

Carbon in the GAC filter is ground and kept loosely inside a cartridge or container. The carbon block filter has a mixture of finely ground powder of AC and food-grade binder, subsequently heated and compressed to form a solid block.

The particles in the carbon block filter are 5–20 times smaller than those in the GAC filter. Either GAC or carbon block filter can be used in PoU water treatment system by having a tradeoff between flow rate and removal of contaminants as GAC filter achieves higher flow rate and lower contaminant removal. In contrast, the carbon block filter has exactly the opposite characteristics. To target specific pollutants, such as lead, and improve performance, additives are mixed with AC to prepare (cast) the carbon block filter.

GAC is extensively used in water treatment to remove organic compounds and residual disinfectants. In this filter, the adsorption of pollutants onto the surface of filter media occurs. Since conventional water treatment plant does not remove DBPs, EDCs or PPCPs, PoU water treatment systems with GAC effectively removes them (Yin & Shang, 2020). This approach effectively removes some inorganic and organic (unwanted taste and odorous compounds, micropollutants) from drinking water. United States Environmental Protection Agency (USEPA) has classified GAC as the best available technology to remove specific organics from water. Chlorine is dosed in the conventional (centralised) water treatment plants to have residual for disinfection while distributing the water to consumers. However, chlorine needs to be removed before treating water in membrane-based PoU water treatment systems as chlorine reacts with the membrane and makes it ineffective for treatment. In a membrane-based PoU water treatment system, pre-treatment is also employed to remove suspended solids to avoid abrasion of membranes. This improves taste and reduces health risks, but it also protects other water treatment system components such as membranes from oxidation or organic fouling.

The following factors are considered for the design of GAC and carbon block filter:

- empty-bed contact time (EBCT)
- hydraulic loading rate
- quantity of AC

Contaminant removal using AC depends on accessibility to the adsorption sites, which is a function of the following parameters. The optimum range of these parameters for improved water treatment is provided below:

- The relative surface area of AC measured by iodine number (IoN) ranges from 900 to 1050.
- Degree of activation of the carbon base measured by carbon tetrachloride number (>50).
- The ability of AC to withstand abrasion is measured by abrasion number (AN). As backwashing or washing with pressurised water may be required in the PoU water treatment system, AN > 70 is required.
- Mesh size of AC. PoU water treatment systems use 8×30 (effective size of 0.8–1.0 mm) or 12×40 mesh size (effective size of 0.5–0.7 mm) of GAC. This helps achieve fast kinetics with an acceptable head loss while covering a large area. In comparison, carbon block filter mesh sizes include 30×200, 50×200 and 80×325, with a typical size of 6.35–25.4 mm block thickness.

- EBCT can be 2 min down to 45 s. EBCT in minutes which is the flow rate through the AC filter

$$EBCT\,(min) = \frac{volume\,of\,media\,(m^3)}{flow\,rate\,(m^3\,min^{-1})} \tag{4.5}$$

- Half-lengths are the period (seconds or minutes) to reduce the concentration of a particular contaminant by 50%. It means about 7.5 lengths are required to remove 99% contaminants and 10 to achieve 99.9% removal.
- Size of AC filter. This can be determined by using volumetric flows in 8020.81–80208.08 L m^{-3} h^{-1}.

4.2.2.1 Dimension of activated carbon filter

A simple concept of EBCT is used for determining the size of the AC filter. EBCT of 45 s removes chlorine from AC (carbon block filter).

$$EBCT\,(s) = \frac{volume\,of\,media\,(m^3)}{flow\,rate\,(m^3\,s^{-1})} \tag{4.6}$$

The flow rate is 30 L h^{-1} or 0.03 m^3 h^{-1}

$$Volume\,of\,media\,(m^3) = \frac{45 \times 0.03 \times 1000}{3600} = 0.375\,L \tag{4.7}$$

This means that a minimum of 0.375 L of AC will be required for removing chlorine from this pre-treatment filter. Because AC filter size is variable, 0.06 m (diameter) $\times$ 0.28 m height is considered, providing about 1 L of AC media volume. This volume of AC media will be sufficient to remove chlorine and, to some extent, chloramine.

$$Volume\,of\,AC\,filter = \frac{\pi}{4} \times 0.06^2 \times 0.28 = 7.92 \times 10^{-4}\,m^3 = 0.792\,L \tag{4.8}$$

Weight of AC = 1.48 kg = ~1.5 kg

An adsorption capacity of 100 mg as chlorine from 1 g of AC was considered in this example. This indicates that 150 g of chlorine can be removed from AC. This example does not consider the desorption of chlorine during the operation of AC pre-treatment units (it is reported that about 15% of chlorine can get desorbed during the operation); thereby increase in the uptake of chlorine is expected during the operation of AC pre-treatment unit.

A maximum concentration of 2 mg L^{-1} of chlorine is considered in feedwater, which should be brought below 1 mg L^{-1} before feeding to the membrane unit. It means 1 mg L^{-1} of chlorine should be effectively removed from AC. Hence, the quantity of water treated from 1.5 kg of AC will be 150 000 L.

If about 100 L day^{-1} of water is dispensed/treated from the PoU water treatment system, AC can last for 1500 days. This indicates that chlorine removal can easily be achieved, and this aspect will not determine the exhaustion of the AC pre-treatment unit.

4.2.3 Membranes (size, length, area, recovery, rejection)

The membrane plays an important role in any PoU water treatment system. Without a membrane, one can say that a PoU water treatment system is not possible. In the RO-based PoU water treatment system, the pressure is applied to a semipermeable membrane, allowing water molecules to pass through while flushing dissolved inorganic substances into the drain. While in the case of UF membranes-based PoU water treatment system, it does not separate water in the same way that an RO membrane does. It is simply an ultra-fine particulate or sediment filter. Mechanical filtration prevents particulates as small as $0.025\,\mu m$ from passing through the UF membrane. For the calculations, we are assuming the PoU membrane module of flux to be $8\,L\,m^{-2}\,h^{-1}$. The membrane module's flux varies from location to location as from feed water, TDS vary.

4.2.3.1 Membrane area calculation

The membrane module used in domestic PoU water treatment systems has a 5.08 cm (2 inch) diameter and 25.6 cm length.

$$\text{Permeate flux} = \frac{\text{Volume of feed water}}{\text{Membrane area} \times \text{time}} \tag{4.9}$$

$$\text{Membrane area} = \frac{\text{Volume of water (per hour)}}{\text{Peremate from membrane} \times \text{time}} \tag{4.10}$$

$$\text{Area of the membrane} = \frac{15}{8 \times 1} = 1.875\,m^2 \tag{4.11}$$

4.2.3.2 Membrane length calculation

The membrane dimension can be calculated as follows:

$$\text{Membrane area} = \text{Length of membrane} \times \text{width of the membrane} \tag{4.12}$$

The width of the membrane is fixed as the length of the module is fixed.

$$\text{Length of membrane} = 7.32\,m \tag{4.13}$$

Hence, it is found that 7.32 m of the membrane is required to produce 15 L of water per hour with a flux of $8\,L\,m^{-2}\,h^{-1}$. Considering the practicality of membrane dimension, an 8 m long membrane may be used in the PoU water treatment system.

The calculation for determining the maximum length of the membrane that can be accommodated in the domestic 7.08 cm (2 inch) RO module is demonstrated below. Assuming the membrane thickness (t_m) is $150\,\mu m$, the permeate tube diameter is 6.35 mm, the feed spacer thickness (t_{fs}) is 0.7 mm and permeate spacer (t_{ps}) is 0.3 mm.

$$\text{Volume of membrane module} = \text{volume of membrane}$$
$$+ \text{volume of feed spacer}$$
$$+ \text{volume of permeate spacer}$$

$$\frac{\pi}{4} \times 0.0508^2 \times w_{\text{module}} = l_{\text{m}} \times t_{\text{m}} \times w_{\text{m}} + l_{\text{fs}} \times t_{\text{fs}} \times w_{\text{fs}}$$

$$+ l_{\text{ps}} \times t_{\text{ps}} \times w_{\text{ps}} + \frac{\pi}{4} \times 0.00635^2 \times w_{\text{pt}}$$

(4.14)

Since widths of the membrane, permeate tube, and spacers are equal, the term w_{module} cancels out in equation (4.14).

$$0.00203 \times w_{\text{module}} = l_{\text{m}} \times t_{\text{m}} \times w_{\text{m}} + l_{\text{fs}} \times t_{\text{fs}} \times w_{\text{fs}} + l_{\text{ps}} \times t_{\text{ps}} \times w_{\text{ps}} + 0.00013 \times w_{\text{pt}}$$

(4.15)

$$l = \frac{0.0019}{(t_{\text{m}} + t_{\text{fs}} + t_{\text{ps}})} = 1.65\,\text{m}$$

(4.16)

Hence, a 2 inch domestic RO module can accommodate a 1.65 m membrane. However, the membrane length is lower in the actual module as there is some tolerance and non-unity packing efficiency.

4.2.3.3 Per cent rejection

Suppose the membrane-based PoU water treatment system is used to treat specific contamination that could damage the consumers' health. In that case, it is helpful to know how much the system can reduce the contaminant in the product. Table 4.1 demonstrates the nominal rejection performance of the two most prevalent types of membranes (CTA and TFC) for various pollutants. Nominal rejection performance for RO membranes at 414 kPa net pressure and 25°C for inorganic and biological and particulate contaminants are shown in Tables 4.1 and 4.2, respectively (WQA Annual education kit).

Due to variations in feed pressure, temperature, water chemistry, pollutant level, net pressure on the membrane, and individual membrane efficiency,

Table 4.1 Inorganic contaminants.

Inorganic Contaminant	Rejection	
	CTA	TFC
Sodium, potassium, bicarbonate, fluoride, chromate, cyanide	85–90%	90–98%
Calcium, cadmium, copper, chromium, magnesium, iron, manganese, aluminium, nickel, zinc, mercury, barium, lead, phosphate, selenium and strontium	90–95%	93–99%
Silver chloride	85–95%	90–98%
Arsenic(V)	85–90%	93–99%
Nitrate	40–50%	85–95%
Phosphate, sulphate	90–95%	93–99%
Boron	30–40%	55–80%
Arsenic^{3+}	60–70%	70–80%

Source: WQA Annual Education Kit.

Table 4.2 Removal of particulate matter and microorganisms.

Biological and Particulate Contaminants	Rejection	
	CTA	TFC
Bacteria, protozoa, amoebic cysts, *Giardia*, asbestos, sediment/turbidity	>99%	>99%
Organic molecules with a molecular weight >300	>90%	

Source: WQA Annual Education Kit.

the actual performance of systems integrating these membranes may be lower. Because the net pressure on the membrane is maximised, countertop RO drinking water systems produce greater overall rejection performance than under-counter systems. While the membrane successfully eliminates iron and manganese, they can quickly foul its surface with deposits even at low concentrations. Other water treatment methods should typically remove iron and manganese before RO treatment. The removal of nitrate is affected by various parameters, including pH, temperature, net pressure across the membrane, and the presence of other pollutants.

While UF, NF and RO membranes are theoretically capable of removing practically all known microbes, including viruses, they cannot guarantee complete safety when used in a household drinking water system. Some pathogens may pass into the treated water due to potential seal leaks and manufacturing flaws. As a result, membrane-based PoU water treatment systems are not normally used as the primary unit/technology for removing microorganisms from a drinking water source. Hence, such systems have UV disinfection after the membrane unit. This multi-barrier approach is adopted to kill/inactivate pathogens if they pass through the membrane due to seal leaks. The degree of rejection of organic molecules with a molecular weight of less than 300 depends on the molecule's size and form. AC is always used in conjunction to ensure the full removal of these smaller molecular weight organic pollutants, preferably as pre-treatment with RO-based water treatment systems.

By comparing the amounts of a specific pollutant in both the feedwater and permeate, the per cent rejection of contaminants for an installed RO membrane may be computed. This calculation can verify actual performance under the customer's unique circumstances.

$$\text{Per cent rejection of contaminant} = \frac{(C_f - C_p)}{C_f} \times 100 \qquad (4.17)$$

where C_f is the concentration of the contaminant in feed water and C_p is the expected concentration of the contaminant in permeate.

4.2.3.3.1 Case study 1: sodium rejection
Sample calculation for contaminant (sodium) concentration in the permeate water is provided below. Assuming the sodium concentration in the feed water

is 500 mg L^{-1}. The concertation of sodium in the permeate can be determined as follows.

For CTA membranes:

$$\text{Concentration of sodium in permeate} = 500 \times \left(1 - \frac{85}{100}\right) \tag{4.18}$$

$$\text{Concentration of sodium in permeate} = 75\,\text{mg L}^{-1} \tag{4.19}$$

For TFC membrane:

$$\text{Concentration of sodium in permeate} = 500 \times \left(1 - \frac{94}{100}\right) \tag{4.20}$$

$$\text{Concentration of sodium in permeate} = 30\,\text{mg L}^{-1} \tag{4.21}$$

No guideline value has been set as per the Guideline for Drinking Water Quality (GDWQ) (WHO, 2022) for sodium in drinking. However, sodium in drinking water of more than 200 mg L^{-1} is not acceptable as it may impart taste to water. Hence, the calculation below shows till what feed concentration of sodium the membranes work satisfactorily to provide potable water.

For CTA membranes:

$$\text{Maximum concentration of sodium in feed} = 1333.33\,\text{mg L}^{-1} \tag{4.22}$$

For TFC membrane:

$$\text{Maximum concentration of sodium in feed} = 3333.33\,\text{mg L}^{-1} \tag{4.23}$$

From the above calculations, it is clear that to bring down the sodium content in permeate (treated water) to an acceptable range, the sodium content in feed should not be more than 1333 and 3333 mg L^{-1} if the feed is being treated with CTA and TFC membrane, respectively.

4.2.3.3.2 Case study 2: fluoride rejection

Sample calculation for contaminant (fluoride) concentration in the permeate water is provided below. Assuming the fluoride concentration in the feed water is 6 mg L^{-1}. The concentration of sodium in the permeate can be determined. The steps are given below.

For CTA membranes:

$$\text{Concentration of fluoride in permeate} = 0.9\,\text{mg L}^{-1} \tag{4.24}$$

For TFC membrane:

$$\text{Concentration of sodium in permeate} = 0.36\,\text{mg L}^{-1} \tag{4.25}$$

For the case of fluoride in water, the guideline value as per GDWQ of WHO (2022) is 1.5 mg L^{-1}. Hence, the calculation below shows the feed concentration of sodium where the membranes work satisfactorily to provide permeate as per the guideline values.

For CTA membranes:

$$\text{Maximum concentration of fluoride in feed} = 10\,\text{mg L}^{-1} \tag{4.26}$$

For TFC membranes:

$$\text{Maximum concentration of fluoride in feed} = 25\,\text{mg L}^{-1} \tag{4.27}$$

From the above calculations, it is clear that to bring down the fluoride content in permeate to an acceptable range, the fluoride content in feed should not be more than 10 and 25 mg L^{-1} if the feed is being treated with CTA and TFC membrane, respectively.

These examples are included to show the design of the membrane for the removal of specific contaminants. They may not be brought down below WHO's guideline values and drinking water acceptable limits depending on other factors such as failure of pre-treatment.

4.2.3.4 Per cent recovery

Percentage recovery measures the amount of water recovered as permeate water. Another way to think about per cent recovery is the amount of water recovered as permeate or product water instead of being discharged as concentrate down the drain. The higher the recovery percentage, the less concentrate water is discarded, and the more permeate water is generated. However, if the recovery per cent is too high for the PoU water treatment system design, scale and fouling might occur, causing more serious issues. The per cent recovery for a PoU water treatment system is calculated using design software, which considers a variety of criteria such as feed water chemistry and pre-treatment before the PoU water treatment system. As a result, the proper per cent recovery at which a PoU water treatment should be operated is determined by its intended purpose.

The formula for calculating per cent recovery is as follows:

$$\text{Per cent recovery} = \frac{\text{Permeate flow rate (LPH)}}{\text{Feed flow rate (LPH)}} \times 100 \tag{4.28}$$

Normally, the PoU water treatment system is designed for a per cent recovery of 50.

$$\text{Feed flow rate (LPH)} = 30 \tag{4.29}$$

The quality of the feed mainly influences the PoU water treatment system recovery rate. Inorganic salts and organics are the most common pollutants, which raise osmotic pressure and foul the system's membranes. For the lesser TDS water, there is a need to apply less pressure to push the water through the pores of the membranes. If too much pressure is applied to the membrane, it may create precipitation of super-saturated salts, or the membranes have a burst pressure limit (8500 kPa) that cannot be overcome.

4.2.4 Membrane module

A membrane module consists of the membranes, a housing, a feed inlet, a concentrate outlet, and a permeate outcome as one unit (Schwinge *et al.*, 2004). In various designs, membrane modules can include hollow fibre, flat sheet, and tubular membranes, as discussed in Chapter 3. Long membrane sheets are bonded together and spirally coiled up around a hollow core tube. A spiral wound membrane or module is the name for this rolled-up arrangement. They are available in various sizes to handle different amounts of water. A household water treatment module may be as small as 2 inch in diameter and 10 inch long, whereas an industrial/community level module may be 4 inch in diameter and 40 inch long. Figure 4.2 shows the nomenclature of membrane modules. It must be noted that only 2 inch modules are used in membrane-based PoU water treatment systems.

The spiral wound membrane modules are the most popular in PoU water treatment systems, owing to their cost-effective packaging, large membrane packing area per element, and comparatively low cost of the materials needed to build the membrane element/module. Membrane flat sheets, feed channel spacers, permeate collecting layers, permeate collection tubes, and sealed carriers (also known as anti-telescoping device – ATD) are typical components of a spiral wrapped RO membrane element available in the market today. ATDs are meant to fit over a membrane element's feed, and concentrate ends. They keep the membrane leaves from elongating ('telescoping') due to a pressure differential across an element. They are also utilised to keep brine seals safe.

The membrane modules consist of the following major non-membrane components of the spiral wound module, as depicted in Figure 4.3:

- feed spacer
- permeate spacer
- permeate tube
- endcap

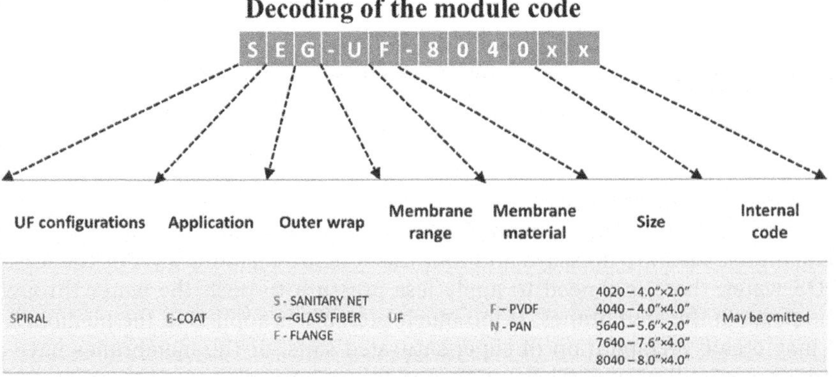

UF configurations	Application	Outer wrap	Membrane range	Membrane material	Size	Internal code
SPIRAL	E-COAT	S - SANITARY NET G –GLASS FIBER F - FLANGE	UF	F – PVDF N - PAN	4020 – 4.0″×2.0″ 4040 – 4.0″×4.0″ 5640 – 5.6″×2.0″ 7640 – 7.6″×4.0″ 8040 – 8.0″×4.0″	May be omitted

Figure 4.2 Nomenclature of the membrane module.

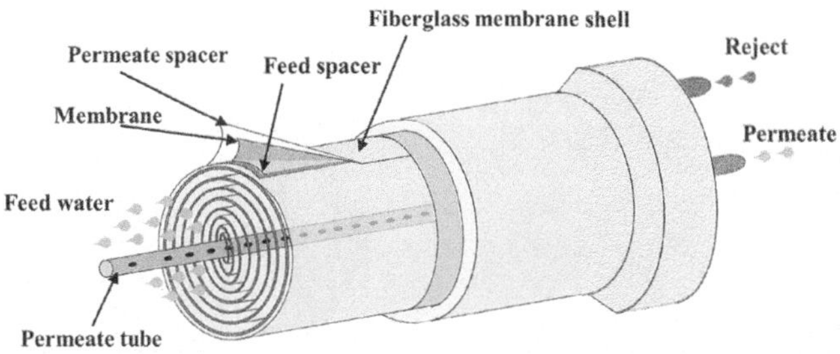

Figure 4.3 Configuration of the spiral wound membrane module.

These elements will be discussed one by one, beginning with a summary of their role and importance. Figure 4.4 shows a spiral wound membrane element without an endcap.

4.2.4.1 Feed spacer
The biplanar extruded net feed spacer is the most frequent feed spacer configuration used in RO membrane modules (Figure 4.5a). Polypropylene

Figure 4.4 Spiral wound membrane element without end cap.

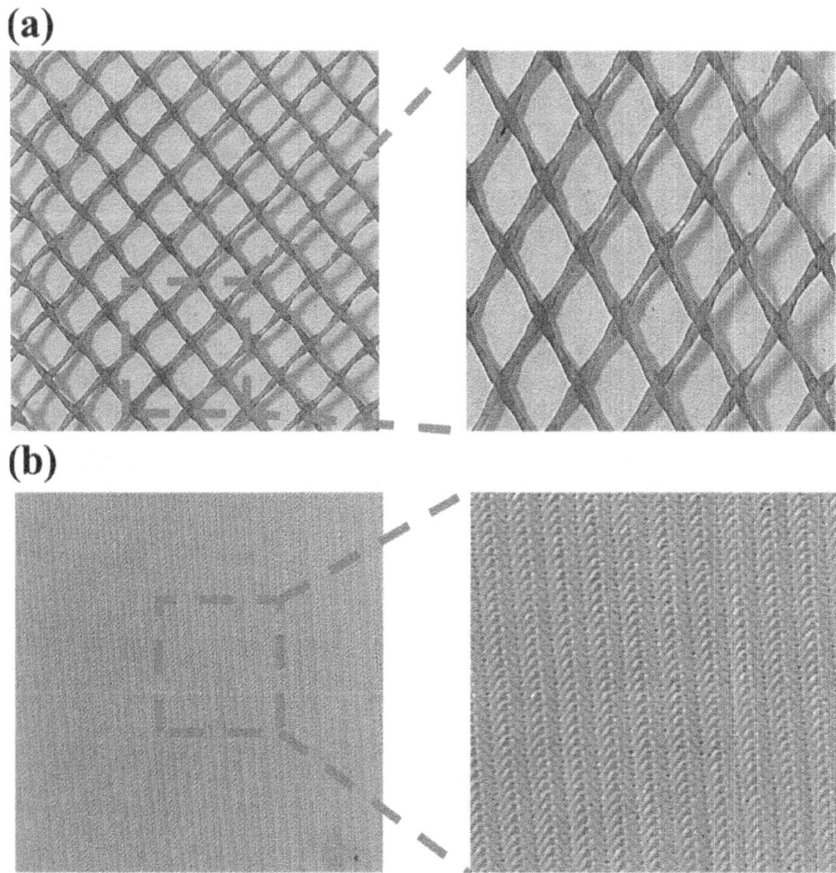

Figure 4.5 (a) Feed spacer and (b) permeate spacer used in RO module.

is used to make most RO feed spacers because it has the best combination of extrudability, low cost, and chemical inertness (Schwinge *et al.*, 2004). Thicknesses of 0.6–0.9 mm are common. The spacer costs less than 1.00 USD m^{-2} for the most regularly used kinds. The spacer is priced below 1.00 USD m^{-2} for the most commonly used varieties.

The feed spacer has two purposes. Ensuring separation between the membrane sheets offers an open pathway for the flowing feed water. It also encourages mixing within the feed channel, which helps move salt and other rejected materials away from the membrane's surface (Kavianipour *et al.*, 2019).

4.2.4.1.1 Maintaining an open feed channel

In producing spiral-wound membrane modules, the key step is rolling up the layered membrane and spacer materials around the permeate tube. The compressive forces exerted during roll-up, and the subsequent tightening of the spiral promote compression of the feed spacer and nesting of nearby feed spacer layers. The apparent thickness of the feed channel, measured after module manufacturing, and the initial thickness of the spacer, taken from a representative sample using a calliper, can be used to determine an apparent change in thickness:

$$\text{Change in thickness}(\%) = \frac{\text{Spacer thickness} - \text{Channel thickness}}{\text{Spacer thickness}} \times 100 \quad (4.30)$$

The apparent channel thickness is obtained by measuring the fabricated module's body diameter and all internal construction elements' thicknesses and lengths (spiral direction). The materials are non-nesting and negligibly compressible, except for the feed spacer, which allows the apparent channel thickness to be estimated theoretically.

The feed spacers used in RO modules are net-type feed spacers that offer contact points with the membrane to sustain and maintain the open feed channel. These spots are created by the intersection of the polymer strands. Each spacer was manufactured into RO modules under the same fabrication procedures, and the difference in thickness was calculated as described above.

This pattern demonstrates a key constraint when it comes to optimising feed spacers. Biplanar extruded nets cannot be altered to the point where they lose their ability to support and separate membrane layers. If the number of intersections is drastically reduced, this can happen. Commercially available spacers for large-scale applications typically have support point densities of $10–12\ \text{cm}^{-2}$.

4.2.4.1.2 Mixing the feed water

The efficacy of spacer mixing, or more precisely, mass transfer, is assessed in terms of the concentration polarisation of a given specie, usually a dissolved salt, that is rejected partially or entirely by the membrane. The polarisation factor (τ) is defined as follows:

$$\tau = \frac{C_{\text{membrane}}}{C_{\text{bulk}}} \quad (4.31)$$

where C_{membrane} is the species concentration over the surface of the membrane, and C_{bulk} is the weighted average concentration (channel cross-section). τ depends on the local permeate flux, mass diffusivity, rejection degree, and mass transfer magnitude.

In the case of sodium chloride, standard spacers and usual operating conditions yield an average of 1.05–1.15. As a result of the improper feed channel mixing, the osmotic barrier in many RO systems is raised by 5–15%. This increases the direct energy consumption in treating very high TDS containing feedwater by 10%. Feed spacers have been proposed to reduce concentration polarisation, although considerable improvements among known configurations result in higher feed channel pressure drop.

4.2.4.1.3 Pressure drop tradeoff

Feed channel pressure loss is a consequence of mechanical support and mass transfer operations. Because membrane modules are often used in multiples in large systems, feed-side pressure drop impacts system performance by lowering trans-membrane pressure and permeating production in downstream modules. Underutilisation led to overutilisation and increased fouling in upstream modules.

Commercial spaces have remained unchanged since their introduction 20 years ago, despite efforts to increase mass transfer by optimising the biplanar extruded net and various forms. The scale of the potential benefit associated with improved mass transfer compared to that gained historically through continuing advancements in membrane chemistry is one of the reasons for this. The low cost of existing spacers is a second consideration. The tradeoff is not impossible, and spacers have been offered as a way to enhance both mass transfer and pressure drop at the same time. Multi-layer spacers, for example, place impediments on the membrane surface where they can effectively interrupt the concentration boundary layer while reducing bulk flow disruption. The use of spacers with non-circular cross-section strands appears to lessen pressure drop while still mixing the boundary layer. Unfortunately, there are no cost-effective large-scale manufacturing methods for such designs.

4.2.4.2 Permeate spacer

Permeate is collected and transported from the membrane to the permeate tube through the permeate spacer. In commercial applications, the most typical spacer is woven polyester cloth (Figure 4.5b). Tricot weaves are popular because of their structural rigidity, smoothness, and fluid-channelling properties (Haidari *et al.*, 2018). The tricot is sandwiched between two sheets of membrane and glued on three edges to form a membrane leaf, as shown in Figure 4.3.

The permeate spacer's pressure drop has a significant impact on module performance. In two ways, the outcome is negative. First, the required net driving pressure to achieve the appropriate permeate flow increases. In other words, the efficiency of the element is diminished. The element efficiency is the ratio of actual permeate flow, Q, to expected output, based on active membrane area (A), membrane permeability (P), and net driving pressure (NDP):

$$\epsilon = \frac{Q}{A \times P \times \mathrm{NDP}} \tag{4.32}$$

Second, the range of variation of the local flux is enhanced for a given average flux within the element. The flow is higher near the root of the leaf, close to the permeate tube. The flux is reduced further away from the tube, near the leaf's tip. As a result, the membrane farthest from the tube may be underutilised, while the membrane closest to the tube may foul early. The shortest range of variation possible is required.

4.2.4.2.1 Permeate spacer pressure drop

The pressure drop within the spacer is almost linear with flow rate and can be parameterised using the following formula:

$$\frac{dP}{dx} = -k\frac{q}{w} \tag{4.33}$$

where dP/dx is the pressure drop in the permeate flow direction at a given distance from the collection tube, q is the volumetric flow rate moving through the spacer at that location, w is the width of the leaf measured parallel to the permeate tube, and k is the friction parameter for the spacer. Due to the squeezing of the woven structure, k varies slightly with applied pressure.

4.2.4.3 Permeate tube

Permeate is collected from the spacer materials inside a module via the permeate tube. The tubes of multi-module pressure vessels are joined in series and act as a conduit for permeate transport to an external manifold. The permeate tube provides critical diagnostic access during operation, allowing conductivity sensors and sample probes to be inserted to look for membrane flaws and leaks. Although materials and tube fabrication procedures have been upgraded throughout the last 20 years of RO module development, tube designs have remained mostly similar. Extruded tubes are commonly used for conventional modules. Side holes and tight-tolerance sealing surfaces are added during secondary machining operations. Injection-moulded tubes are sometimes used for shorter modules.

4.2.4.4 Endcap

Endcap design and functioning have received renewed attention in recent years. The endcap is a precision-engineered, injection-moulded plastic component that serves several functions in the module. The following is a list of some of those roles:

- **Leaf retention**: The endcap, also known as an ATD, stops the membrane leaves from telescoping (relative axial movement).
- **Load transmission**: The endcaps transfer axial load from module to module and into the module's robust fibreglass cover.
- **Bypass prevention**: The endcap contains a brine seal that prevents feed water from bypassing the module and entering the annulus between the module and the pressure vessel's interior wall. The connection between the fibreglass shell and the endcap aids in preventing brine seal bypass.
- **Permeate connection**: The endcap has been developed in some situations to contain interlocking and permeate sealing elements between modules.

4.2.4.4.1 Interlocking end caps

Sliding couplers, such as the one illustrated in Figure 4.6, have been utilised to connect the permeate tubes of adjacent spiral wound membrane modules in pressure vessels for more than 20 years. Although membrane providers offer a

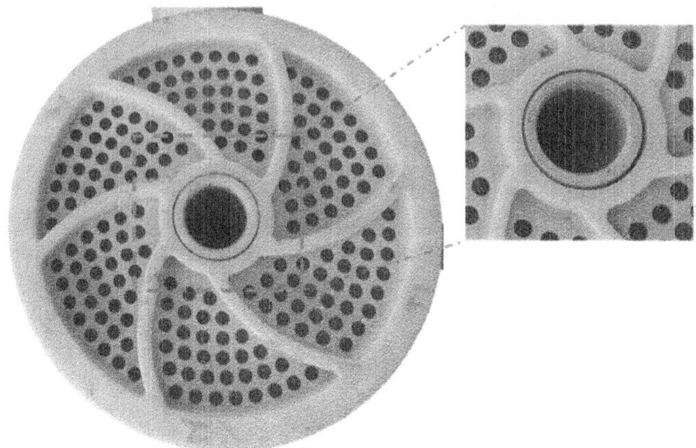

Figure 4.6 Standard sliding coupler is used to connect the permeate tubes of adjacent elements.

variety of coupler designs, they all work on the same principle: a pipe segment with radially compressed O-rings on both ends, coupled to the adjacent permeate tubes either internally or externally. For increased durability and to eliminate sliding coupler issues, the coupler's sealing functions were shifted to the endcap in the form of a standard O-ring face seal. The components' rotating locking compresses the permeate face seal, removing the potential of incorrect coupler installation and subsequent seal wear.

4.2.5 Post-treatment
It is essential to treat the permeate after it has been collected to avoid corrosion of downstream piping and equipment, as the treated water is devoid of dissolved solids. The RO product water is slightly acidic (pH 5–6), soft, and low in alkalinity, whereas distilled water is near-zero hardness and has alkalinity, unlikely to exceed 2 mg L^{-1}. These waters corrode metal and asbestos cement pipes and absorb calcium from mortar-lined pipes. Depending on the raw water chemistry, post-treatment for low-pressure membranes (MF and UF) is typically minimal, consisting of disinfection (as a secondary barrier) and, in some cases, pH adjustment and corrosion control.

4.2.5.1 Chlorination
The final step in water treatment is post-chlorination. It introduces a low level of chlorine into the water supply from the point of origin to the point of distribution. Its goal is to keep residual chlorine at a level to provide disinfection potential against pathogens in stored water. However, chlorination is not practised in plumb-in PoUs as membranes and UV disinfection (if provided) impart sufficient protection against pathogens.

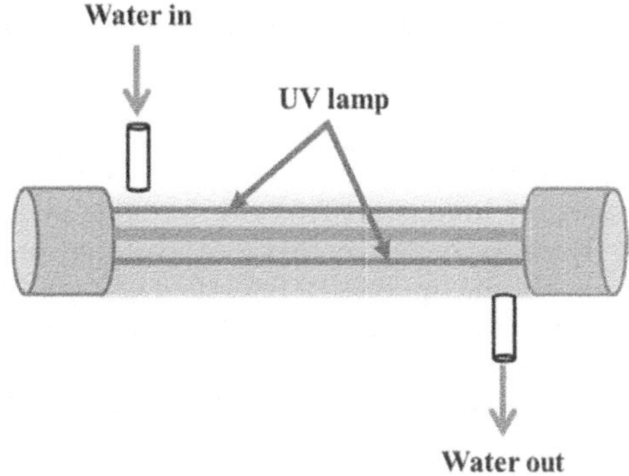

Figure 4.7 Small-scale UV disinfection system.

4.2.5.2 *Ultraviolet disinfection*

UV treatment (Figure 4.7) is a widely used approach for disinfection water. UV disinfection is a cost-effective and safe water treatment method. UV radiation is used to inactivate bacteria, viruses, moulds, algae, and other pathogens in water. The optimum wavelength is 250–270 nm. Most manufacturers provide a lamp intensity of 30 000–50 000 μ W s cm^{-2}. In general, coliform bacteria, for example, are destroyed at 7000 μ W s cm^{-2}. Based on the vendor sales literature, most of the commercially available lamp tube thickness is 16 mm.

There are following two types of UV lamps that can be used Chatterley and Linden (2010):

- **Low-pressure lamps**: A low-pressure mercury arc lamp resembling a fluorescent lamp produces UV light (monochromatic) in 254 nm. Standard lengths of low-pressure lamps are 0.75–1.5 m with 1.5–2.0 cm diameters. The ideal lamp wall temperature is between 95 and 122°F. Conversion to UV-C is typically 30–35%.
- **Medium and high-pressure lamps**: Medium pressure UV lamps function at much higher pressures, temperatures and power levels and emit a broad spectrum of higher UV energy between 200 and 320 nm. The disinfection capacity is much faster. However, for UV disinfection in the PoU water treatment system, low-pressure lamps and systems are perfectly suitable and even preferable to medium-pressure lamps and systems due to their lower power consumption.

4.2.5.3 *Total dissolved solids adjustment/controller (bypass, adding chemicals, etc.)*

As mentioned earlier, RO membranes remove TDS and the output water is almost devoid of TDS. This remains a major limitation of the RO membrane-based water treatment system. Opponents of RO have always contested the use of RO and its possible health impacts due to the consumption of low TDS water. Hence, manufacturers altered the configuration of the RO membrane-based

PoU water treatment system and introduced arrangements to make up for the lost dissolved solids. In one of the arrangements, the TDS control screw manually adjusts the flow of water from the two water inputs to control the TDS levels in the output (treated) water. When TDS levels need to be raised, water flows through the RO membrane decreases and the flow through the AC filter or UF membrane increases. This can be done by simply turning the TDS control screw. Because heavy metals like lead and arsenic can only be removed by an RO membrane, mixing RO purified water with non-RO filtered water can result in many heavy metals in the treated water. Another option for maintaining dissolved solids in output water is to add chemicals in appropriate doses depending on the quantity of water to be treated. These chemicals mainly consist of calcium and magnesium compounds. The most common substances used are calcite (calcium carbonate) and corosex (a magnesium compound).

Figure 4.8 shows the mass balance in the TDS control unit in a typical PoU water treatment system.

$$\dot{m}_d \times TDS_d = \dot{m}_f \times TDS_f + \dot{m}_p \times TDS_p \tag{4.34}$$

where m is the mass flow rate, and p, f, and d represent the permeate from RO membrane-based PoU water treatment system, a fraction of pre-treated feed water and finally treated water.

The following example indicates steps to estimate TDS levels in finally treated water by blending permeate with raw water:

Let us assume the rejection of TDS from membrane-based PoU water treatment system is 94%, and the feed water has a TDS level of 1200 mg L^{-1}; therefore, the TDS level in the permeate water will be

$$TDS_p = 1200 \times (1 - 0.94) = 72 \, mg \, L^{-1} \tag{4.35}$$

The mass flow rate from the membrane-based PoU water treatment system will be fixed (depending upon the size of the membrane module), and the flow rate from the system is assumed as 20 L h^{-1}. Now suppose we require 15 L h^{-1} water with a TDS level of 300 mg L^{-1}, applying mass balance for the TDS control unit

$$\dot{m}_d \times TDS_d = \dot{m}_f \times TDS_f + \dot{m}_p \times TDS_p \tag{4.36}$$

$$\dot{m}_f = \frac{4500 - 20 \times 72}{1200} = 2.55 \, L \, h^{-1} \tag{4.37}$$

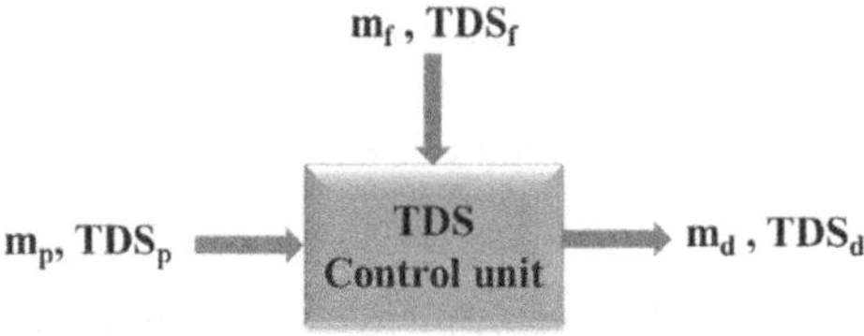

Figure 4.8 Mass balance for TDS control in PoU system.

Therefore, we need to add 2.55 L h^{-1} of pre-treated feed water to permeate water to get the desired TDS levels in finally treated water. This also means that the blending of raw water should be carried out carefully, as almost 10% of raw water is added to maintain TDS levels. Raw water quality should also be carefully analysed to determine the percentage of raw water added to permeate. Otherwise, the concentration of critical water quality parameters may exceed the guideline values/standards after blending a higher proportion of raw water to permeate.

4.3 DESIGN OF MULTI-STAGE MEMBRANE-BASED POINT-OF-USE WATER TREATMENT SYSTEM

Multi-stage PoU water treatment system has a different connotation and can be classified as (1) multiple water treatment filters, for example UF and RO membranes, (2) multiple stages of the same type membrane, for example having only more than one RO membrane (Medeiros *et al.*, 2020). While multi-stage filters having different types of membranes are fairly common, a multi-stage system with the same type membrane is still at the exploratory stage of commercialisation due to increased cost and space requirements (Figure 4.9).

Water recovery rates are higher in multi-stage membrane-based PoU water treatment systems than single-stage systems. Instead of draining the concentrate, it travels through extra membrane elements, which produce more permeate from the same amount of total feed water entering the system. When optimising a multi-stage membrane-based PoU water treatment system for high recovery rates, it is critical to follow the manufacturer's specifications for system pressure and flow requirements. The membrane elements, pressure vessels, pressure pumps, and pipe configuration must be properly engineered to fulfil the pressure and flow requirements. Additional pumps may be required, increasing the energy consumption of the system. Furthermore, because the concentrate from multi-stage systems has higher amounts of dissolved solids

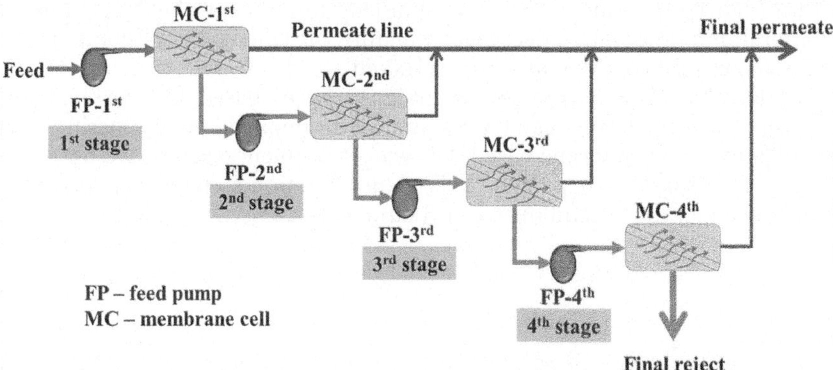

Figure 4.9 Schematic diagram of the multi-stage RO system used in this study.

and pollutants, disposal or treatment of the concentrate may differ from that of single-stage systems. When considering the possible reduction in reject water due to implementing a multi-stage water treatment system, these additional costs should be addressed.

In addition, a high proportion of reject water (>50%) is also a significant drawback of RO-based PoU water treatment system. Due to increased awareness about water conservation, consumers and critiques are demanding increased recovery of water from the RO-based water treatment system. Multi-stage PoU water treatment system is probably the only practical option for increasing the recovery rate. This will increase the cost of the PoU water treatment system besides the space requirement for installation of the system.

4.4 SUMMARY

Membrane-based PoU water treatment systems have evolved rapidly over the last half-century, from laboratory discovery to plants capable of producing up to half a million daily tonnes of desalinated saltwater. This transition has undergone significant changes in membrane chemistry, module design, and RO plant configuration and operation.

Configuration of membrane-based PoU water treatment systems is almost optimum due to the availability of many variants in the market. There is not enough literature on designing these systems due to the system's proprietary nature. It appears that most of the manufacturers are keeping design data strictly confidential. Moreover, enough experience is already obtained in the optimised configuration of these systems and going back to design criteria is no longer required. Moreover, the necessity of obtaining design criteria for these systems is minimum because specifications/sizing of these systems for almost all the capacity are already available. Considering the resilience of membrane-based PoU water treatment systems in removing almost everything from water has reduced the requirement for developing/having design criteria for water quality parameters. Design criteria for hydraulics are also available for large plants, which can also work for these systems. These systems are designed with a higher safety factor. They are unlikely to fail even due to significant deterioration in the source (feed) water quality, provided proper operation and maintenance of these systems are undertaken.

We have attempted to develop design criteria based on the available literature, manufacturers' catalogues and, more importantly, discussing with manufacturers of membrane-based PoU water treatment systems. We have also utilised design criteria available for large desalination systems to explain the sizing of membrane-based PoU water treatment systems.

doi: 10.2166/9781789062724_0105

Chapter 5

Modelling membrane operations in membrane-based point-of-use water treatment systems

5.1 INTRODUCTION

Globalisation, sustainability, collaboration, invention, discoveries, and evolution are the driving factors of today's globe. Under the restrictions imposed by the notion of sustainable development, meeting the rising demand for raw materials, energy, and goods is a difficult task. Mathematical modelling is a powerful technique for generating correct understanding and related issues for scaling up. It delivers immediate insights into membrane system parameters such as flow, fouling, and resistance build-up. As a result, simulating membrane operations will enhance the PoU water treatment system's performance. While trying to understand phenomena like fouling and pressure losses, the results of membrane scales are frequently deceptive. Flat sheet membrane scale experiments can produce either under or overly predictive real-world scaled-up results. More often than not, this can lead to incorrect expectations, resulting in erroneous encouragement or discouragement.

In the literature, three types of mathematical modelling are commonly encountered. The first is transport process modelling, which entails first-principles-based models and results simulation. Membrane engineers have been using this method for a very long time. This method can solve a wide range of systems, from the most basic to the most complex. The first principle-based modelling strategy has proven versatile for understanding membrane separation from liquid filtration to gas permeation. The classical thermodynamics technique is the second sort of modelling approach. This method is particularly effective for modelling processes such as phase inversion and pore development in the manufacture of polymer membranes. Thermodynamics also aids in the knowledge of entropy formation and consequently related irreversibility in processes, which provides a hint of possible mitigation measures. The thermodynamic analysis also aids in determining the viability of operations and, as a result, provides insight into how membrane technology might improve efficiency. The third modelling approach is newer and has grown in popularity over time as computers and robust algorithms for solving non-linear fluid flow equations become available. This type of modelling is known as computational fluid dynamics (CFD), and it is now widely employed by membrane technocrats.

CFD is now used in membrane module design, packing efficiency computations, flow phenomena analysis, and other previously uncharted fields. The first principle modelling and thermodynamic modelling provided insight into the divergence from theoretical limits, and ideas for how to construct systems to achieve the lowest desalination energies began to emerge. CFD flow modelling in commercial modules and module design has been utilised to achieve improved hydrodynamic flow patterns, resulting in less fouling and higher flux.

With the advancement of new membrane technologies, mathematical modelling is becoming increasingly important in making them industrially and economically viable. As a result, membrane-based solutions or 'ideas' that appear infeasible now could be a solution many decades down the road. As a result, any technological advancement in this field is critical for our progeny and the species' survival. In membrane operations for water treatment applications, process modelling is extremely significant. It allows users to get precise forecast results, benefiting membrane-based water treatment plants from design to commissioning, implementation, and maintenance. Application of optimisation principles and CFD models to PoU systems is limited, and downscaling of desalination plants remains a major tool for designing these systems. This chapter will introduce and discuss mathematical modelling techniques of membrane operation in membrane-based PoU water treatment systems.

5.2 GENERAL PRINCIPLES OF MODELLING

Mathematical modelling enables the basics of membrane separation phenomena to be understood during the water treatment process. The correct membrane design equation can be used to determine the water treatment system size and the membrane area required to meet the desired output. Furthermore, mathematical modelling can be integrated into the control system to predict the membrane system's fouling potential, enabling the implementation of appropriate measures to decrease and mitigate membrane fouling. As a result, it can be stated that the operation and maintenance of membrane-based water treatment units can be greatly improved with the use of appropriate predictive modelling for the membrane process.

To simulate a fluid flow, CFD employs a numerical method. This method is semi-empirical, and the temperatures and concentrations predicted are consistent. Furthermore, because these equations are intended for a certain geometry and flow rate regime, the models cannot be utilised to optimise geometry consistently.

Various numerical and analytical transport models have been developed to understand membrane-based processes better. Three modelling methodologies have been used in general,

- one-dimensional (1-D) or
- two-dimensional (2-D) or
- three-dimensional (3-D) models.

The first strategy considers membrane modules only in 1-D for process modelling, resulting in an oversimplifying calculation. As a result, the problem's precision was compromised. 1-D mathematical modelling and empirical correlation are used to design or monitor a process. However, these approaches do not account for oversimplified geometries under simplified settings. They cannot assess axial flow field fluctuations or concentration profiles along a flow channel. The second technique uses modelling tools to investigate process specifics using 2-D or 3-D model simulations. Within the module, it is possible to provide detailed information regarding transport phenomena while considering the effect of geometrical variations on process performance. Compared to 3-D simulations, the 2-D modelling technique still loses the specifics of the process. Models in three dimensions provide further evidence by solving conservation equations simultaneously while considering the geometric variations of the process.

In conjunction with the CFD community of models, these methodologies are used in various engineering disciplines. They have resulted in the continual improvement of several processes, yielding robust solutions to specific engineering challenges. The local distribution of parameters like pressure, temperature, species, and velocity for various geometries can be monitored in a model. As a result, realistic models with optimised assumptions and hydrodynamic interactions can aid in the resolution of specific problems.

5.3 TRANSPORT MODELS USED IN MODELLING MEMBRANE PROCESSES

Membrane modelling is critical because it allows users to gain useful information, particularly about a membrane's ability to retain certain solutes while enabling the permeability of specific substances. The mechanism of membrane penetration and rejection is a complex process. As a result, a good predictive model allows the performance of the membrane to be precisely forecasted without the need for time-consuming and complicated procedures to gather raw data for prediction and improve the efficiency of the process by optimising it. A more accurate model reduces the number of experiments and, as a result, lowers costs and time. In the membrane process, modelling is typically used to anticipate two aspects of performance: flux prediction and rejection prediction. Several models have been proposed to explain how water molecules move through a porous membrane.

Different transport processes are used for different types of membranes. Because an RO membrane is dense, constituents (solute) are transported through it by solution diffusion. In contrast, transport processes for microporous membranes like MF and UF are determined by the pore size of the membranes. The behaviour of NF membranes is similar to that of dense and microporous membranes. Furthermore, size exclusion and electrostatic interaction are the two key factors influencing how RO/NF rejects solutes. In the water treatment process, the transport mechanism predicts membrane performance (flux and rejection). Figure 5.1 depicts a simplified classification of transport models for the four membranes addressed in this chapter and the criteria influencing the type of water treatment models employed.

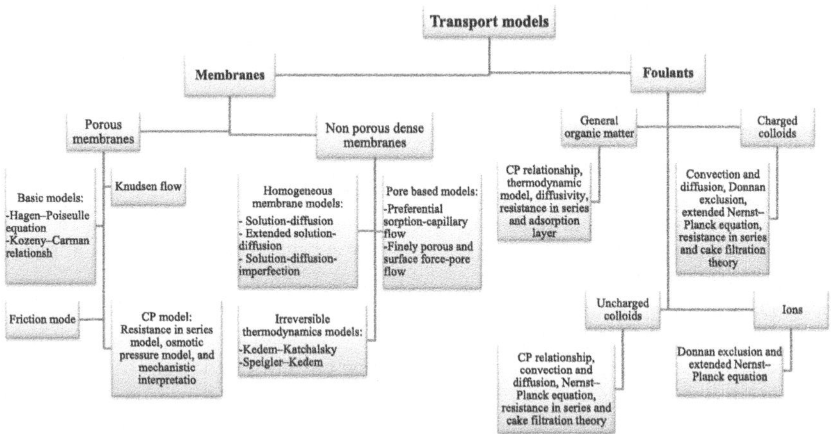

Figure 5.1 Transport models classification based on membrane and foulants type.

From Figure 5.1, it can be observed that the membranes utilised and the foulants in the raw water affect the transport models used to anticipate and explain the water treatment process. Membranes are further divided into porous and non-porous forms, each with its own set of mathematical modelling equations and theories. The characteristics of foulants determine which model is most appropriate and accurate in defining the retention mechanism. In transportation systems, several principles are applied, including size exclusion and charge or dielectric exclusion. The sieving effect, also known as steric hindrance, is the primary rejection mechanism for uncharged solutes. Solutes with a molecular weight greater than the membrane's cutoff are kept. Convection owing to pressure differences and diffusion through a concentration gradient across the membrane also contribute to the transfer of uncharged solutes. The Donnan effect involves the interaction between co-ion rejection and the fixed electric charges associated with the membrane matrix and is the main mode of transport for charged solutes (ions). The next sections will go through some of the most often used transportation models. It should be highlighted that no mathematical modelling has yet been developed that applies to a wide range of applications. The majority of existing models are only usable in select and limited circumstances.

The solution-diffusion model is the most generally accepted transport model for water molecules flowing over a polymeric membrane. This model's theory is built on a few key assumptions. First, it is assumed that all permeable compounds dissolve in the polymer matrix and permeate through it and that the membrane is devoid of holes. Although it is difficult to say that a membrane is completely devoid of pores, the pores in question appear and disappear as time passes and water diffuses through the membrane. The second assumption is that the surface of the membrane and the fluids on each side of it are in a condition of equilibrium. The researcher can sense a continual chemical potential gradient

Table 5.1 Basic transport models for modelling membrane processes.

Transport Model	Basic Features/Equations
Stefan–Maxwell model: Teorell–Meyer–Sievers (TMS) model by Hassan *et al.* (2007)	• Defines electrical properties of membrane in terms of effective charge density and electrostatic effects. • Uniform radial distribution is assumed. $$\sigma_{\text{salt}} = 1 - \frac{2}{(2\alpha - 1)\aleph + (\aleph^2 + 4)^{0.5}}$$ $$P_{\text{salt}} = D_s(1 - \sigma_{\text{salt}})\left(\frac{A_k}{\Delta x}\right)$$
Stefan–Maxwell model by Peppas and Meadows (1983); Roberson and Zydney (1988); Basile *et al.* (2015)	$$J_s = -D_2 C x_s \frac{d\ln(y_s x_s)}{dz} + \frac{D_2}{D_1} J_v C x_s$$ x_s is the mole fraction of the solute; y is the activity coefficient of the solute; C is the total molar concentration; and D_1 and D_2 are the overall transfer coefficients are given in terms of the binary Stefan–Maxwell diffusivities
Hydrodynamic or pore model by Pappenheimer (1953)	• Ion transport described by the Nernst–Planck equation. • Ion concentration and radial distribution of electric potential described by the non-linear Poisson–Boltzmann equation, force balance in narrow pores relative to pore length described by Naiver–Stokes equation. $$J_s^d = -D_O f(q) S_D A_k \frac{dC}{dx} - D_{\text{eff}} A_k \frac{dC}{dx}$$ S_D is the steric factor restricting, $f(q)$ is the viscous drag at the wall, D_O and D_{eff} are the diffusion coefficients in a free solution and pore, respectively, A_k corresponds to the membrane surface porosity $$S_D = (1 - q)^2 f(q) = \frac{1}{(1 + 2.4q)}$$

(*Continued*)

Table 5.1 Basic transport models for modelling membrane processes (*Continued*).

Transport Model	Basic Features/Equations
Hydrodynamic or pore model by Renkin (1954)	Solute flux in terms of diffusion flow and convection flow: $$J_s = D_O(q)S_D A_k (C_m - C_f) + f(q)S_c J_v C_m$$ Restriction factor S_c for the convection flow is introduced: $$S_c = 2(1-q)^2 - (1-q)^4$$ $$f(q) = 1 - 2.104q + 2.09q^3 - 0.95q^5$$
Hydrodynamic or pore model by Verniory *et al.* (1973)	Modified classical pore models by adopting wall correction factor and relating it to the treatment with thermodynamic and frictional forces: $$J_s = D_O f(q)S_D A_k \Delta C_s / \Delta x + g(q)S_c J_v \overline{C_s}$$ Introduced the convective friction forces function, $g(q)$ in addition to $$f(g):f(q) = \left(1 - 2.105\,q^2 + 2.865\,q^3 - 1.7068\,q^5 + 0.72603\right)/\left(1 - 0.75857\,q^5\right)$$ $$g(q) = \left(1 - 2/3q^2 - 0.20217q^5\right)/\left(1 - 0.75857q^5\right)$$
Kedem and Katchalsky (KK) model (irreversible thermodynamics) (Kedem and Katchalsky, 1958)	Describes flow of solution and solute, including performance measured using the dissipation function. $$J_v = L_p\left(\Delta P - \sigma\Delta\pi\right)$$ $$J_s = P_s\left(c_m - c_p\right) + (1-\sigma)J_v$$

Kedem–Katchalsky–Zelman model (irreversible thermodynamics) (Nair *et al.*, 2018)	For multi-solute system. Considers differential equations for multiple solute transport.
Spiegler–Kedem model (irreversible thermodynamics) (Nair *et al.*, 2018)	Takes into account the variation in concentration profile at high flux and concentration gradients. Membrane transport is characterised by solvent, solute permeability and reflection coefficient Based on the average concentration inside the membrane Commonly utilised in single-solute systems $$J_s = -P'\left(\frac{dc}{dx}\right) + (1-\sigma)J_v$$
Extended Spiegler–Kedem model (irreversible thermodynamics) (Nair *et al.*, 2018)	A 1-D model for multiple solutes. Includes solute–solute interactions Characterised by parameters obtained from experimental data in the literature using the Levenberg–Marquardt approach in conjunction with the Gauss–Newton algorithm
Steric hindrance pore (SHP) model (air *et al.*, 2018)	Wall correction factors are eliminated and used to predict: pore radius and membrane porosity to thickness ratio Adapted for permeation of a single solute (neutral), but not suitable for salt retention application $$J_i = v_i k_i \left[H_\mathrm{F}, i u_x c - H_\mathrm{D}, i D_i \left(\frac{dc}{dx} + C \frac{z_i F \, d\Phi}{RT \, dx} \right) \right]$$ H_F and H_D are the steric hindrance and frictional force's parameters that impede convective and diffusive transport, respectively

thanks to this assumption. Finally, the pressure in the membrane is assumed to be constant, resulting in a thermodynamic chemical potential gradient across the membrane solely determined by the concentration gradient. Assuming that concentration and pressure gradients alone govern water flow, the relationship for change in chemical potential is expressed as

$$d\mu_i = RTd \ln a_i + v_i dP \tag{5.1}$$

where a_i is the activity, v_i is the molar volume, and dP is the pressure gradient across the membrane and i is the ith component.

Several hypotheses and models have been presented to arrive at an appropriate model for membrane transport mechanisms, as shown in Table 5.1. Due to the differences in membrane properties, the models utilised for each membrane may differ. The complication of the models is also affected by various types of fouling.

5.4 MODELLING REVERSE OSMOSIS PROCESS

Several mathematical models have been presented for the solute and solvent transport mechanisms, with the solution-diffusion transport mechanism being the most popular and commonly used. Fick's law can express the water transport through the membrane. The equation is reduced to a formula in which the water flux is connected to pressure and concentration gradients across the membrane after various assumptions and derivations. The salt content at the feed across the membrane and on the permeate side determines solute flux.

Pre-processing, solution, and post-processing are the three basic phases of most commercial CFD software. The first stage in constructing a meaningful model geometry is pre-processing. In the PoU system, the spiral wound membrane module is the most usually employed. While modelling the spiral wound membrane module, it is critical to consider the pressure drop.

The flow in the feed and permeate channels can be quantitatively represented by establishing a set of partial differential equations, often known as the Navier–Stokes equations. The following are the governing equations for mass and momentum conservation in laminar flow:

$$\frac{\partial p}{\partial t} + \nabla.\left(\rho\vec{\mu}\right) = S_j \tag{5.2}$$

$$\rho\frac{\partial v}{\partial t} = -\nabla P + \nabla.\left(\mu\nabla\vec{\mu}\right) + \rho g + S_\mathrm{m} \tag{5.3}$$

where ρ is the fluid density, v is the velocity, $\vec{\mu}$ is the fluid velocity vector, μ is the viscosity of the fluid, P is the pressure, and g is the acceleration due to gravity. S_j and S_m are the mass and momentum source terms, respectively. Because of penetration through the membrane wall, source terms are required for accounting for losses or gains at permeable wall boundaries. The following equation can be used to incorporate the conservation of solute (A) mass fraction into the model:

$$\frac{\partial \rho m_A}{\partial t} + \nabla.\left(\rho\left(\vec{\mu}\right)m_A\right) - \nabla.\left(\rho D m_A\right) = 0 \tag{5.4}$$

where m_A is the solute mass fraction and D is the solute mass diffusivity coefficient.

For laminar flow steady-state conditions, the transport of solute (A) in terms of solute concentration can be defined using the convection–diffusion equation given as

$$\nabla.\left(D_A \nabla \vec{\mu}\right) - \nabla.\left(\vec{\mu} c_A\right) = 0 \tag{5.5}$$

where D_A is the solute diffusion coefficient and c_A is the solute molar concentration of (A).

Depending on a situation's characteristics and solution method, 2-D or 3-D conservation equations are applied. The governing transport equation with no time-dependent components is used for the feed and permeate channels in steady-state laminar flow. This model can examine the effects of operating pressure, solute concentration, and cross-flow velocity on total and local fluxes throughout the length of the membrane module. A distinct source term must be included in each transport equation to describe the changes in mass or momentum on both sides of the membrane. The following are the source terms for the continuity equation, momentum transfer equation, and energy equation (Roy *et al.*, 2020):

$$S_j = \begin{cases} -\dfrac{J.A}{V} \text{ at the feed/membrane interface} \\[2ex] \dfrac{J.A}{V} \text{ at the permeate/membrane interface} \end{cases} \tag{5.6}$$

$$S_m = \begin{cases} -\dfrac{J.A.\mu}{V} \text{ at the feed/membrane interface} \\[2ex] \dfrac{J.A.\mu}{V} \text{ at the permeate/membrane interface} \end{cases} \tag{5.7}$$

$$S_k = \begin{cases} -\dfrac{J.A.\mu\left(c_p \Delta T + h\right)}{V} \text{ at the feed/membrane interface} \\[2ex] \dfrac{J.A.\mu\left(c_p \Delta T + h\right)}{V} \text{ at the permeate/membrane interface} \end{cases} \tag{5.8}$$

where μ is the velocity of the feed stream, V is the volume of the fluid element, and h is the latent heat. At the feed/membrane contact, the mass, momentum, and energy sink terms can be calculated. As shown in Figure 5.2, the data are concurrently sent to the membrane/permeate interface as correct and relevant results.

5.4.1 Modelling transport phenomena in reverse osmosis membrane
The solvent molecules (water) migrate through the membrane from the high concentration side to the low concentration side during an osmotic process

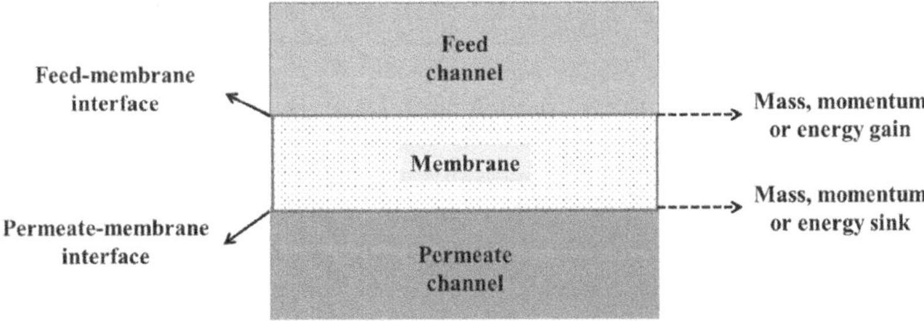

Figure 5.2 Mass, jump approach to calculate mass, momentum and energy source term: S_j, S_m, S_k.

because of the difference in osmotic pressure. The transport mechanism in the RO process works in the opposite direction of the osmotic process. As a result, high pressure is provided to the feed solution side (which must be greater than the osmotic pressure) to compel water molecules to flow to the dilute (permeate) side across a dense hydrophilic membrane (Figure 5.3).

RO membrane's design equations are based on the solution-diffusion transport mechanism. The flux of water and solutes is calculated as follows:

$$J_w = P_w \left(\Delta p - \Delta \pi \right) \tag{5.9}$$

$$J_s = P_s \left(C_1 - C_2 \right) \tag{5.10}$$

The osmotic pressure of the liquid feed is one factor to consider when determining the operating pressure for RO. The pressure provided to avoid the inward flow of water over the membrane is called osmotic pressure. This

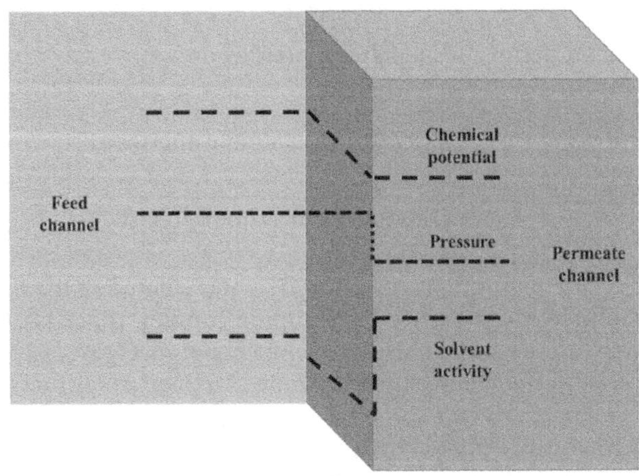

Figure 5.3 Chemical potential, pressure and solvent activity difference between feed and permeate during RO process.

is proportional to the solute concentration and temperature for low solute concentrations (Ang and Mohammad, 2015). The Pitzer equation or Van't Hoff's reported equation can be used to compute osmotic pressure. The Pitzer equation's computation stages, on the other hand, are difficult. So here is a simplified version of Van't Hoff's equation:

$$\pi = \frac{n}{V_{\mathrm{m}}} R_{\mathrm{G}} T \tag{5.11}$$

It should be noted that this equation is only applicable to dilute water solutions with a TDS of less than 20 000. The existing transport models and basic equations for estimating the solute and solvent fluxes for RO membranes are summarised in Table 5.2. The reference given has the details of the derivation.

Table 5.2 Merits and demerits of the models and/or predicting organic trace rejection by RO membranes.

Model	Merits	Demerits
Models based on irreversible thermodynamics	• No specified mechanism of solute transport and structure for membrane. • Appropriate for performance prediction of multiple solutes (NF/RO systems).	• The driving forces such as pressure and concentration gradients restricted real-world application of the model. • Series of assumptions leads to a case to unrealistic cases. • For the model applicability, system should not be near equilibrium. • Effective for high rejection system.
Mass transport models	• Simple models: provide estimates even for technically demanding separations. • Model linearisation results in faster calculations.	• Different water quality and operating conditions affect solute mass transfer coefficients. • Membrane physicochemical parameters constrain model application. • Highly appropriate to single-solute systems. • Mass transfer coefficients of the solute depend on the test unit scale in membrane scale-up.
Artificial neural network models in conjunction (or not) with quantitative structure–activity relationship models	• Easy to use. • Do not use physical principles or transport phenomena, hence overcoming complexity difficulties. • Accurate estimates than existing models. • Model is valid regardless of the membrane's rejection performance.	• Applicable in specific range of experimental conditions. • Model accuracy is influenced by changes in membrane characteristics due to fouling or swelling.

Mathematical modelling of trace contaminant rejection is extremely valuable in forecasting the removal of harmful compounds from water. The irreversible thermodynamics model proposed by Kedem and Katchalsky (1958) is another model that describes boron transport. The solution-diffusion and irreversible thermodynamics models' transport equations are shown above and in Table 5.1.

The water temperature in membrane water treatment processes is important to consider when functioning the plants. Higher temperature raw water necessitates lower operating pressure, and vice versa for low-temperature raw water to achieve the specified production capacity. The permeate to flow at 25°C can be approximated by assuming all membrane performance factors remain constant (Ang and Mohammad, 2015):

$$Q_T = Q_{25°C} \times 1.03^{T-25} \tag{5.12}$$

Both RO and NF membranes can be used with this equation. The selectivity of RO membranes is often measured as the solute rejection coefficient, which is calculated as (Ang and Mohammad, 2015):

$$R = \frac{C_b - C_p}{C_b} \times 100\% \tag{5.13}$$

Because RO membranes reject a significant amount of solute, the solute concentration maintained at the membrane surface (feed side) is very high. Concentration polarisation (CP) becomes a problem as a result of this. The solute concentration at the membrane surface (feed side) differs from that in bulk. As a result, the solute wall concentration is calculated using the film theory model (as shown in the previous section). To calculate the RO membrane's true rejection percentage, the solute wall concentration is required.

After the CP influence is taken into account, the governing equations for water and solute flux are also modified. The final calculation provided in the previous part might be used to calculate the solute concentration at the membrane surface.

For rejection:

$$R_{CP} = \frac{C_m - C_p}{C_m} \times 100\% \tag{5.14}$$

For more accurate prediction, the CP effect should be included in modelling. The following is an example of how the CP ratio for tubular membranes in the turbulent area has been calculated:

$$\frac{C_m}{C_b} = \frac{1}{D_r} + \left[1 - \frac{1}{D_r}\right]\exp(k) \tag{5.15}$$

$$D_r = \frac{C_b}{C_p} \tag{5.16}$$

The following equation can be obtained by including the CP ratio in the equations (5.9) and (5.10):

$$J_{w} = k_{w}\left[\Delta p - \Delta \pi \left(\frac{C_{m}}{C_{b}}\right) + \frac{\Delta \pi}{D_{r}}\right] \qquad (5.17)$$

$$J_{s} = k_{i}C_{b}\left(\frac{C_{m}}{C_{b}} - \frac{1}{D_{r}}\right) \qquad (5.18)$$

These equations allow for the estimation of RO membrane performance and the prediction of the effect of CP on the given operating circumstances.

Turbulent flow problems require a different procedure than laminar flow problems resulting in a wider range of length and period. Because of the turbulent eddies in the flow, turbulent flow issues are inherently unsteady and 3-D. This makes it more difficult to model turbulent flow equations and more computationally demanding to solve them. The Reynolds averaged Navier–Stokes (RANS), direct numerical simulation, and large eddy simulation approaches have all been utilised to solve turbulence problems. The direct numerical simulation (DNS) solves the Navier–Stokes equations for all turbulent flow scales under unstable conditions without using a turbulence model. However, DNS computations for turbulent flows become complex when the Reynolds number in the flow domain grows. As a result, DNS modes need highly precise 3-D mesh manipulation and enormous computational resources. However, because most PoU systems are small, they may be simplified, and laminar flow can be considered via membranes, which can then be studied.

5.4.2 Membrane reliability modelling

The examination of the reliability of a membrane process unit should be an integral aspect of its design and operation. However, because this has typically been difficult to assess and quantify, its management has been generally overlooked. Throughout the life cycle of the membrane module, reliability modelling can be utilised to predict failure and subsequent process shortages. This would also result in considerable capital and operating cost reductions by deploying superior strategies drawn from simulation results while allowing utilities and plant designers to effectively control and quantify the risk of non-compliance due to membrane failure. The application of reliability modelling expands membrane process performance beyond membrane fouling to include membrane ageing and failure.

Modelling membrane processes accurately is crucial for evaluating innovative process configurations, building scalable membrane systems, estimating process costs, and directing future research. Most membrane process models sacrifice accuracy for computational efficiency by using simplified process approximations and solution parameters. For RO case examples, this approach measures the error introduced by these simplifications. While the error level caused by these simulations varies depending on the case study characteristics and specifications, the model formulations can underestimate or overestimate average water flux by two times. While the magnitude of error introduced by the simulations varies depending on the parameters and specifications, the model formulas can underestimate or overestimate membrane processes operating under standard conditions.

These performance models may contain errors that mask high-impact technology development research demands, impede technological scale-up from the lab, and allow persistent research in non-competitive technologies. Detailed 1-D process models link the design, operational, condition, and process elements of a membrane stage using a system of differential equations. When these models are solved for a specific design and operational condition, the profiles of variables along the membrane stage are described. It provides an estimate of how long the process will take. Other important parameters like net energy consumption and cost can be evaluated and optimised using stage-level process models, which can then be integrated into systems-scale models.

5.4.3 Modelling transport and fouling mechanisms

Figure 5.4 depicts the transport of solute/solvent molecules across membranes. There are four types of transport mechanisms in general:

- bulk flow
- diffusion
- restricted diffusion
- solution-diffusion

Fouling affects the membrane flux and rejection of undesirable substances in water. As a result, modelling membrane performance in water treatment should include membrane fouling as an inescapable phenomenon. Fouling may exist indifferent forms:

- **Adsorption**: Occurs when the membrane interacts with the foulant/solute in the solution in a specific way.
- **Pore blockage**: Foulants/particles obstruct the pores of the membrane.
- **Deposition**: A layer-by-layer accumulation of particles on the membrane's surface (known as cake resistance fouling as well).
- **Gel formation**: In the immediate proximity of the membrane surface, CP causes the formation of a gel layer.

Fouling mechanisms, phenomenological background, and transport equations in RO membranes are all detailed in Table 5.3.

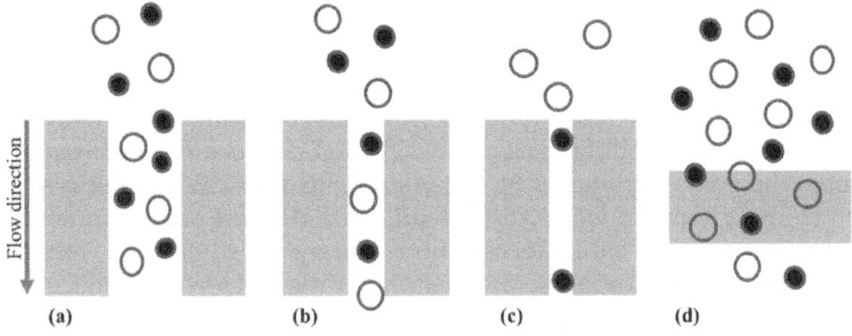

Figure 5.4 Transport mechanisms in membranes: (a) bulk flow; (b) diffusion; (c) restricted diffusion; (d) solution-diffusion.

Table 5.3 Fouling mechanisms, phenomenological background and transport equations.

Fouling Mechanism	Phenomenological Background	Flux Equation for Dead-end	Flux Equation for Cross-flow
Completely blocked pores	Large particles (greater than pore size) completely block pores	$J = J_0 \exp(-K_b t)$	$J = \left(J_0 - J^*\right)\exp\left(-k_2 t\right) + J^*$
Internal pore blockage	Small particles (smaller than pore radius) enter the pores or absorbed or deposited on the pore walls	$J = J_0\left[1 + \dfrac{1}{2} \cdot K_s \cdot (A \cdot J_0)^{0.5} \cdot t\right]^{-2}$	$J = J_0\left[1 + \dfrac{1}{2} \cdot K_s \cdot (A \cdot J_0)^{0.5} \cdot t\right]^{-2}$
Particle pore blocking	Particles on surface may seal, bridge, or partially block pores, or attach to inactive regions	$J = J_0\left(1 + K_i \cdot (A \cdot J_0) \cdot t\right)^{-1}$	$J = \dfrac{J^*}{\left[1 - \left(\left(J_0 - J^*\right)/\left(J_0\right)\right)\exp\left(-J^* k_1 t\right)\right]}$
Cake filtration	Particles do not enter the pores or seal the pores but form a cake layer on the membrane surface	$J = J_0\left(1 + 2 \cdot K_c \cdot (A \cdot J_0)^2 \cdot t\right)^{-\frac{1}{2}}$	$k_0 t = \dfrac{1}{J^{*2}}\left[\ln\left(\dfrac{J}{J_0} \cdot \dfrac{\left(J_0 - J^*\right)}{\left(J - J^*\right)}\right) - J^*\left(\dfrac{1}{J} - \dfrac{1}{J_0}\right)\right]$

Source: Ang and Mohammad (2015).

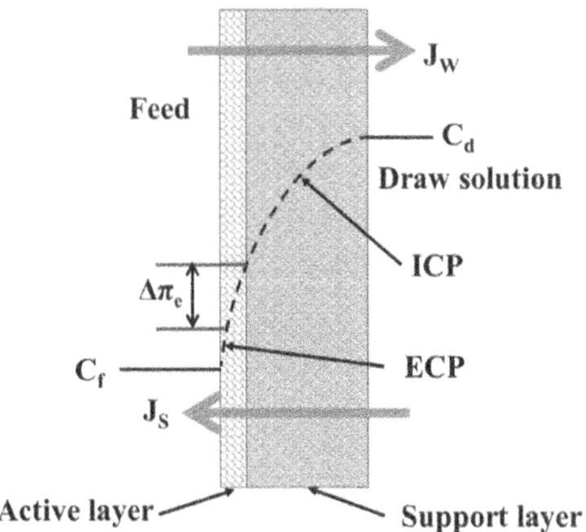

Figure 5.5 Concentration polarisation.

5.4.4 Modelling concentration polarisation

The aggregation of residual solutes in the membrane boundaries at the feed side is called concentration polarisation (CP) (Figure 5.5). It is an unavoidable consequence of membrane selectivity. Because the solute concentration on the membrane filtration unit's wall differs from the bulk feed concentration, CP makes modelling the membrane filtration unit more difficult. The concentration of solutes on the solute wall is difficult to determine. This phenomenon has been described using the boundary layer film model. The convection of solute over the membrane surface is equal to the solute that penetrates through the membrane layer under steady-state conditions, with solute diffusion returning to the bulk feed solution.

$$JC = JC_p - D_{ji}\frac{dC}{dz} \tag{5.19}$$

The solution of Eq. (5.18) for the boundary conditions: $z = 0; C = C_m$ and $z = l_{bl}; C = C_b$ will give

$$J = \frac{D_{ij}}{l_{bl}}\ln\left(\frac{C_m - C_p}{C_b - C_p}\right) \tag{5.20}$$

Finally, the solute concentration at the membrane surface is determined using this equation derived from the film theory model.

$$C_m = \left(C_b - C_p\right)\exp\left(\frac{Jl_{bl}}{D_{ij}}\right) \tag{5.21}$$

C_m is a very significant parameter as it has been used in several membrane transport models.

Usually, the term $\left(D_{ij}/l_{\mathrm{bl}}\right)$ in Eqs (5.19) and (5.20) is called a mass transfer coefficient $k_{i,\mathrm{b}}$, and it can be estimated using some standard dimensionless numbers such as the Sherwood number (Sh), Reynolds number (Re), Schmidt number (Sc), and Peclet number (Pe), the analytical tools can be used to determine the bulk and permeate concentrations.

A membrane purification unit's effectiveness may be hindered by the presence of CP. It reduces water flux, rejects unwanted solutes, produces precipitation due to high surface concentrations exceeding the solubility limit, alters membrane separation properties, and increases fouling from colloidal or particle debris that clogs the membrane surface. As a result, the membrane module's design and operating circumstances are critical in preventing and reducing the impact of CP.

5.4.5 Energy consumption

One of the major components of membrane-based PoU water treatment is energy consumption. The energy required for the PoU system can be expressed as

$$E_{\mathrm{T}} = E_{\mathrm{in}} + E_{\mathrm{pt}} + E_{\mathrm{p}} + E_{\mathrm{A}} \tag{5.22}$$

where E is the energy requirement, subscripts T, in, pt, p, and A are the total, feed water supply, pre-treatment and post-treatment, pressure pump and accessories (chemical dosing, filter backwashing/cleaning and pumping the product water).

It is generally believed that the process of RO material transfer is affected by external and internal resistance. The internal resistance is mainly the osmotic pressure that must be overcome during the RO process, and the external resistance is the thermodynamic limit pressure. As the manufacturing technology of RO membrane elements has become increasingly mature in recent years, the thermodynamic limit resistance exceeding the transmembrane transmission resistance has become the dominant factor limiting the flux of the RO membranes.

5.4.5.1 Specific energy consumption

Specific energy consumption (SEC) is the most important parameter to characterise the performance of the RO desalination process. The unit is kWh m^{-3}, representing the energy required to produce 1 m^3 of product water. It is also the stable operation of RO desalination equipment. The main influencing factors of specific energy consumption include the osmotic pressure of concentrated water, the resistance of the fluid through the membrane, the friction loss in the concentrated water and freshwater channels of the membrane, and the efficiency of the pump.

Figure 5.6 shows a schematic diagram of an RO membrane-based PoU water treatment process. As a result of the main analysis of the influence of the transmembrane transfer resistance and the thermodynamic limit resistance on the membrane flux, the mathematical model is simplified, and the effect of CP is ignored. The pressure drops of the concentrated water flowing out of the RO membrane are ignored, and the permeate water and the feed water pressures are considered equal.

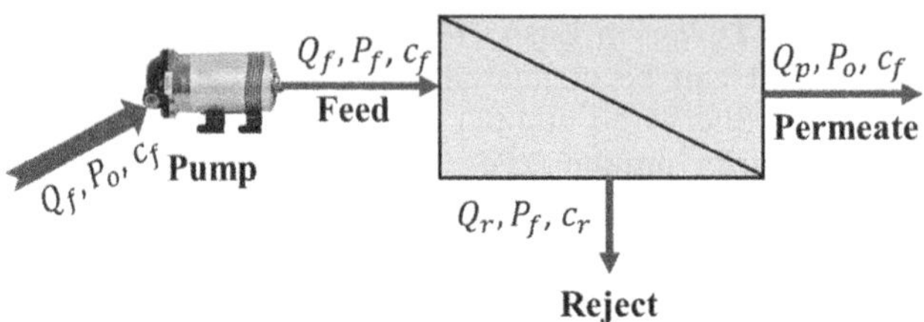

Figure 5.6 Energy flow diagram of the RO process.

The pressure pump raises the feed pressure from P_0 to P_f to overcome the osmotic pressure. The efficiency of the pressure pump is expressed by η_{HP}, and the energy consumption W of the high-pressure pump can be expressed as

$$W = \frac{Q_f \times (P_f - P_0)}{\eta} \tag{5.22}$$

The SEC can be defined as

$$\text{SEC} = \frac{W}{Q_P} = \frac{\Delta P_f}{Y_\eta} \tag{5.23}$$

In the entire process, according to the law of conservation of materials, there is a mass balance equation for solution and solute; that is, the feed water flow is equal to the sum of the freshwater flow and the concentrated water flow:

$$Q_f = Q_p + Q_r \tag{5.24}$$

where Q_f is the feed water flow; Q_P is the permeate flow and Q_r is the concentrate flow. Similarly, according to the conservation of salt quality, the salt content in the feed water is equal to the sum of the salt content of the permeate water and the salt content of the concentrated water, namely

$$Q_f c_f = Q_p c_p + Q_r c_r \tag{5.25}$$

where c_f is the average salt concentration of the feed water; c_p is the average salt concentration of the osmotic water; c_r is the average concentration of salt in the concentrated water.

The osmotic pressure of the solution can be calculated according to the Van't Hoff equation as described in earlier sections.

The osmotic pressure π_{exit} at the concentrated water outlet of the membrane element has the following relationship with the original seawater osmotic pressure π_0:

$$\frac{\pi_{exit}}{\pi_0} = \frac{c_r}{c_f} = \frac{1 - Y(1 - R_s)}{1 - Y} \pi_0 \tag{5.26}$$

where R_s is the desalination rate.

The osmotic pressure π_{p} of the permeated water is ignored; that is, the osmotic pressure difference at the outlet of the membrane element is

$$\Delta\pi_{\mathrm{exit}} = \pi_{\mathrm{exit}} - \pi_{\mathrm{p}} \cong \frac{1 - Y(1 - R_{\mathrm{s}})}{1 - Y}\,\pi_0 \tag{5.27}$$

Assuming that the salt concentration is the highest at the concentrated water outlet of the membrane element, the transmembrane pressure entering the membrane element must be greater than the osmotic pressure difference at the membrane outlet to ensure that the membrane element has permeation flux over the entire length, and the flux is only in the membrane. The exit of the component is reduced to 0.

$$\Delta P_{\mathrm{f}} \geq \Delta\pi_{\mathrm{exit}} \tag{5.28}$$

Equation (5.25) is the thermodynamic limit equation of the RO process in cross-flow operation. The permeate water production process can occur on the entire RO membrane element until the exit of the membrane element tail, where the effective driving force disappears:

$$\mathrm{SEC} \geq \frac{1 - Y(1 - R_{\mathrm{s}})}{Y(1 - Y)\,\eta_{\mathrm{HP}}}\,\pi_0 \tag{5.29}$$

Standardising the SEC relative to the feed water's osmotic pressure (π_0) can be obtained as follows:

$$\mathrm{SEC}_{\mathrm{norm}} = \frac{\mathrm{SEC}}{\pi_0} \geq \frac{1 - Y(1 - R_{\mathrm{s}})}{Y(1 - Y)\,\eta_{\mathrm{HP}}}\,\pi_0 \tag{5.30}$$

5.4.5.2 Energy efficiency

In addition to reducing the specific energy consumption of the membrane process, improving the energy utilisation efficiency of the RO process can also extend the working time of the small RO seawater desalination equipment. Therefore, a mathematical model is established in this section to calculate the membrane energy efficiency, permeate volume flow and desalination rate in a small system, which is mainly related to the feed water concentration and the high-pressure pump.

Due to the irreversible nature of the mixing process, the separation process of the membrane does not occur spontaneously. To carry out the process, the pressure difference applied by the pump must take the form:

$$\Delta P_{\mathrm{pump}} = \Delta\pi + \Delta P_{\mathrm{losses}} \tag{5.31}$$

where $\Delta\pi$ is the osmotic pressure difference between the membrane inlet and outlet solution; $\Delta P_{\mathrm{losses}}$ is the friction loss in the membrane system.

When the membrane loss in the RO system is not considered, that is, $\Delta P_{\mathrm{losses}} = \Delta\pi$, the system consumes the least energy to complete the membrane separation work in an ideal state. According to the second law of thermodynamics, considering the adiabatic mixing process, the minimum separation energy of the feed liquid salt and water can be obtained from the following equation:

$$\omega_{\min} = -RT\left(y_\omega \ln\left(y_\omega\right) + y_s \ln\left(y_s\right)\right)$$

(5.32)

where y_s is the molar fraction of the solute in the membrane inlet solution, that is, the ratio of the amount of solute to the amount of the entire solution; and y_ω is the mole fraction of the solvent, $y_\omega = 1 - y_s$. Therefore, the minimum separation work $W_{\min}$ of the membrane can be expressed as

$$W_{\min} = \rho_f Q_f \omega_{\min}$$

(5.33)

where ρ_f is the density of the feed solution.

W_{pump} represents the actual work provided by the pump to the system due to friction loss, which can be obtained by

$$W_{\text{pump}} = \frac{Q_f\left(P_f - P_0\right)}{\eta_p}$$

(5.34)

The energy utilisation efficiency of the membrane is defined as the ratio between the minimum separation work and the work done by the actual pump:

$$\eta = \frac{W_{\min}}{W_{\text{pump}}}$$

(5.35)

5.5 CASE STUDIES

Application of mathematical models, e.g. CFD in optimising the design of membrane-based PoU water treatment system, is rarely cited. There are several reasons for not having examples of the application of mathematical models in such systems a) CFD is complex and the efforts/resources required to optimise these systems are more than other tools such as actual systems; b) PoU water treatment systems are comparative cheaper options and any optimisation may not result in effective cost/resources reduction c) experimentation is carried out using actual systems for further optimisation. However, mathematical models have different advantages and should be applied in optimising even PoU water treatment systems.

An attempt is made to apply computational models using examples in optimisation and further improve the performance of PoU water treatment systems.

5.5.1 Modelling velocity field and concentration polarisation

This case study uses a 2-D streamline upwind Petrov/Galerkin finite element model to simulate the spiral wound RO module (Ma *et al.* 2016). The model uses Navier–Stokes equations and solute transport equations. It must be noted that this model is not relevant to ultrafiltration and microfiltration modules because the flow characteristics in these systems are typically turbulent or transitional, which this model cannot adequately represent. However, the model can accurately replicate concentration polarisation in most spiral wound RO systems and some NF systems under normal operating conditions.

5.5.1.1 Velocity field

The parabolic velocity profile in the empty feed channel regions is identical to that of a channel with impermeable walls. However, due to penetration, the primary flow changes from cross-flow to flow towards the membrane surface in the thin layer near the membrane surface (Figure 5.7). The contours of flow velocity magnitude in a channel with a single 0.5 mm × 0.5 mm filament affixed to the membrane surface are shown in Figure 5.8. Compared to the parabolic

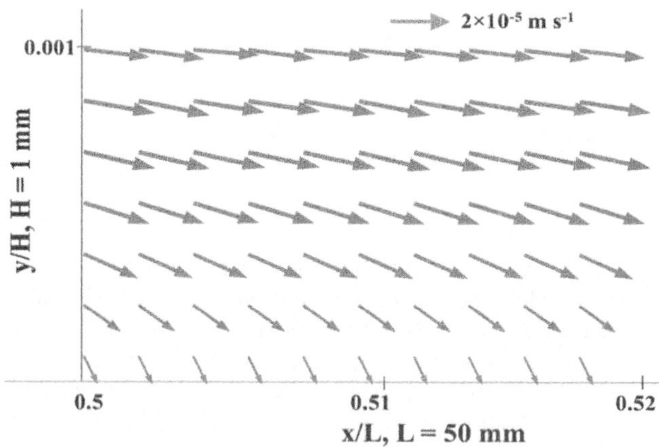

Figure 5.7 Velocity field in the flow direction transition region in an empty channel.

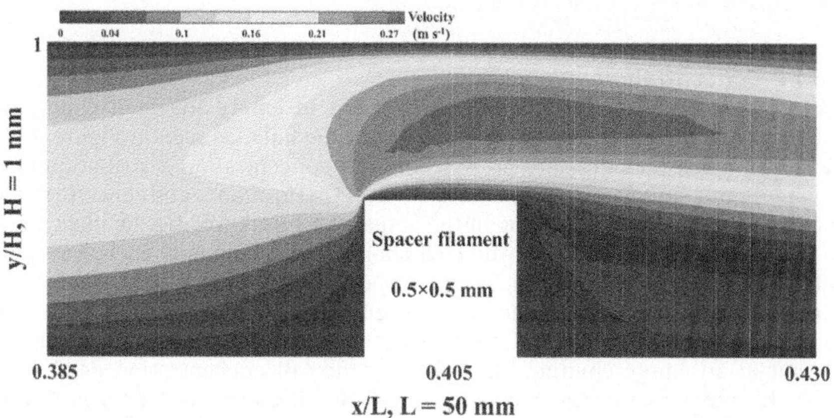

Figure 5.8 Velocity profile contour in a feed channel with a 0.5 mm × 0.5 mm filament attached to a membrane at $y=0$.

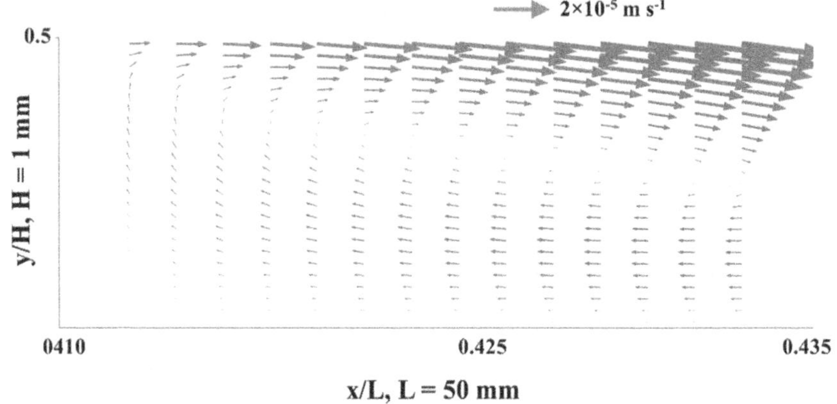

Figure 5.9 Velocity profile in the recirculation regions in a feed channel with a 0.5 mm × 0.5 mm filament attached to a membrane at $y = 0$.

velocity profile, the velocity in the restricted feed channel surrounding the filament increases dramatically, implying that mass transfer in these locations would be enhanced. However, if the filament was linked to the membrane, the velocity dropped in the regions directly in front of and behind the filament near the membrane surface; mass transfer would be compromised in these areas. The attached filament formed recirculation zones (wakes), primarily behind the filament. Figure 5.9 depicts the recirculation induced by a 0.5 mm × 0.5 mm filament attached to a single membrane. Because mesh length (the distance between two nearby filaments) is normally at least 5–10 times spacer filament thickness in practical RO modules, contact between adjacent filaments may be negligible under normal working conditions.

5.5.1.2 Concentration polarisation

Salt content grew monotonously downstream in empty RO feed channels, then decreased from the membrane surface to the bulk, as seen in Figure 5.10. However, when filaments are present, the salt concentration distribution can become highly complex. The prediction of system performance and concentration polarisation (CP) understanding in the actual systems, knowing the intricacies of the concentration profile in the feed channel with filaments is critical. The salt content profile in a channel section is depicted in Figure 5.11. Under the same operating conditions, the salt concentration in the low-velocity zones directly in front of and behind the filament increased significantly in contrast to that in an empty channel; nevertheless, the salt concentration decreased near the other membrane facing the filament. Because wall concentration decreases and increases in different regions in the affected area compared to an empty channel, the effect of the filaments on the overall recovery rate is complex, i.e., the recovery rate may increase, decrease, or remain unchanged

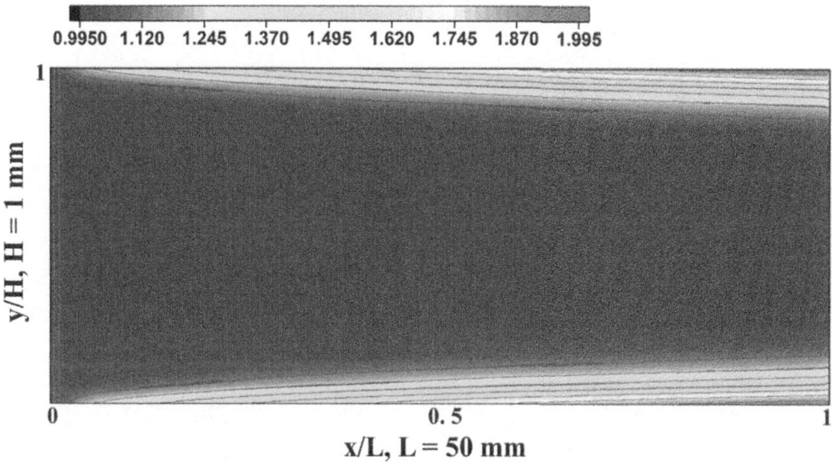

Figure 5.10 Salt concentration profiles in an empty feed channel.

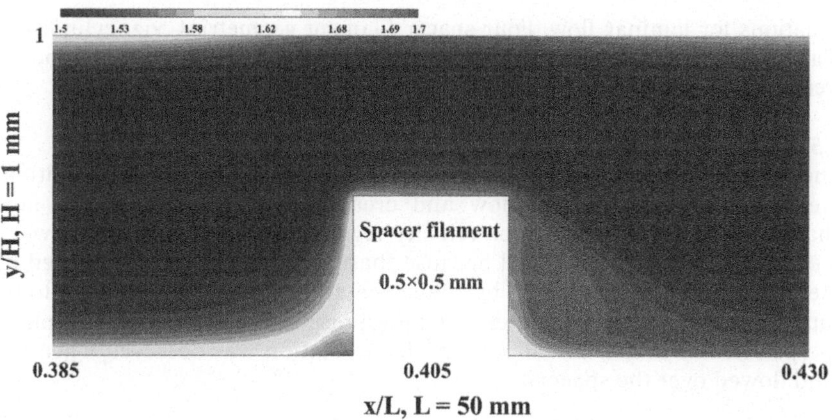

Figure 5.11 Salt concentration profiles in a feed channel with a 0.5 mm × 0.5 mm filament attached to a membrane at $y=0$.

when compared to an empty channel. The outcome is determined by various operating conditions, feed channel and filament configurations, and membrane characteristics.

5.5.2 Effect of spacer filament geometry on velocity field and concentration polarisation

This case study analyses the effect of different spacer filament geometry on the velocity and field concentration polarisation in RO module (Singh *et al.*, 2022). The model combines mass transfer in the membrane with 3D Navier-Stokes

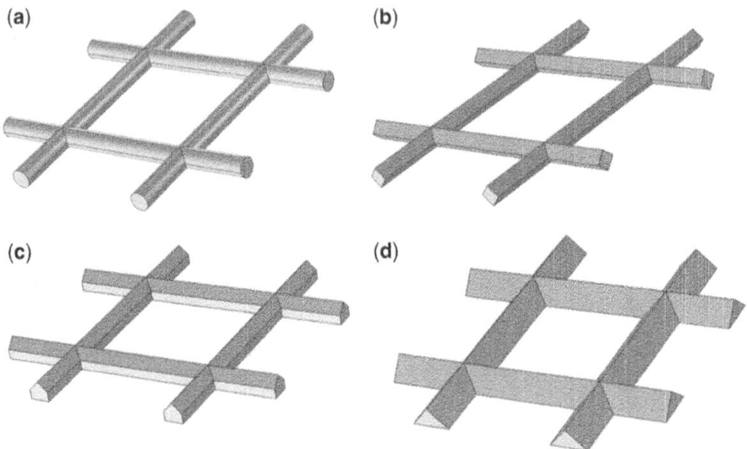

Figure 5.12 Different spacer filament geometry (a) cylindrical (b) diamond (c) pentagonal (d) triangular.

equations for laminar flow. Four spacer filament geometries, viz. cylindrical, diamond, pentagonal and triangular, available as commercial net-type spacers, were used for the study (Figure 5.12).

5.5.2.1 *Velocity field*

The presence of spacers in the RO module produced an intermittent breakdown of the laminar flow and created a disturbance in the flow channel (Figure 5.13). A higher velocity region is present in space between the spacer and membrane wall because these regions are mostly affected by the fluid acceleration created by the spacers. Velocity magnitude contours superimposed with streamlines are also presented at a chosen z-x plane (perpendicular to the flow direction). Recirculation was observed when the fluid flowed over the spacers.

5.5.2.2 *Concentration polarisation*

The concentration distribution for the different spacer geometry-filled RO modules is shown in Figure 5.14. CP can be reduced by introducing the spacer in the channel because the spacer produces the fluid acceleration effect, increasing the shear stress near the wall. Among different spacer configurations, the channel containing cylindrical spacer filaments showed the highest CP. This was attributed to the low fluid acceleration due to the smooth cylindrical shape of spacer filaments, as discussed in the previous section. The channel containing triangular spacer filament showed the least CP due to enhanced fluid mixing and acceleration. The concentration boundary layer thickness was higher at the bottom side of the channel containing triangular spacer geometry. This was due to the triangular filament's flat base, which could not enhance mixing. Similar observations were made in earlier studies.

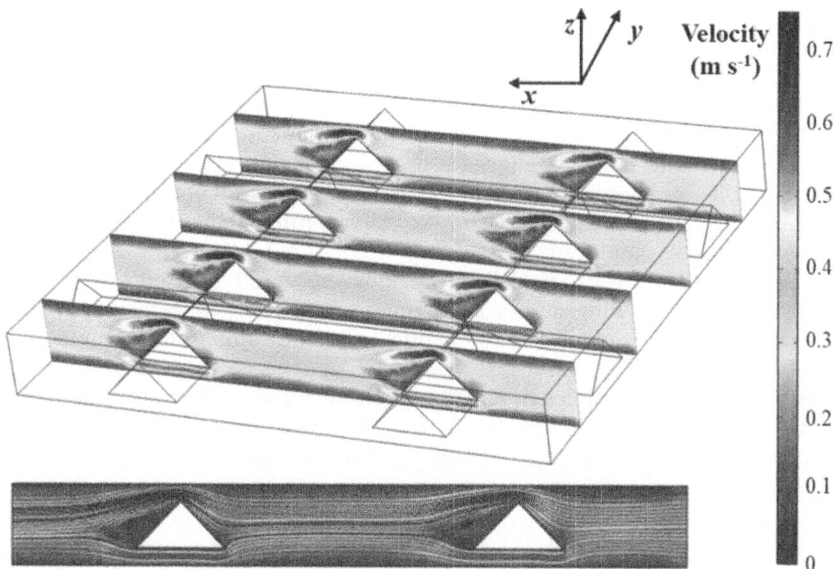

Figure 5.13 Velocity field and streamline for triangular spacer.

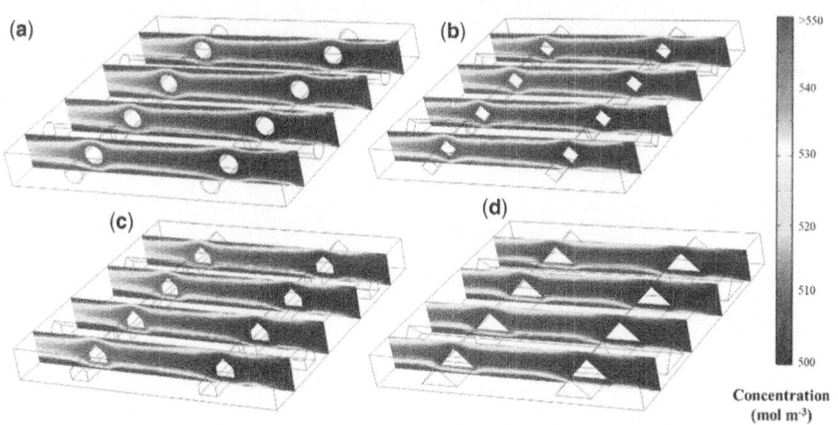

Figure 5.14 Concentration contour for different spacer filament geometry (a) cylindrical (b) diamond (c) pentagonal (d) triangular.

5.6 SUMMARY

Numerical simulation is one of the most efficient methods to study CP in real systems and optimise channel/module design. Mathematical models, particularly CFD, are widely employed in membrane-based water treatment plants. CFD flow modelling is used in module design to achieve optimisation of hydrodynamic flow patterns to have higher flux from these systems. The application of CFD models to PoU water treatment systems is limited for various reasons. The mechanism of solution traversing membrane and rejection of solute is a complex phenomenon. Hence, a precise development and application of the model can precisely forecast the performance of the membrane, and time-intensive and complex alternate processes of optimising the system can be avoided. This chapter presents a simplified classification of transport models for the four types of membranes and the criteria influencing the type of water treatment models employed.

The solution-diffusion model is the most accepted transport model for water molecules flowing across a polymeric membrane. Several hypotheses and models are presented in this chapter to select an appropriate model for membrane transport mechanisms. The models applied for each membrane may differ due to the differences in membrane characteristics. The outcome of the models are used in developing velocity field, concentration field and CP and simulation results were compared with experimental data.

doi: 10.2166/9781789062724_0131

Chapter 6
Operation and maintenance of membrane-based point-of-use water treatment systems

6.1 INTRODUCTION

Sustainability or unhindered continuation of water treatment systems is primarily driven by interactions among factors linked to the environment, technology, and consumers. Although several examples of unsustainable community and household-based water treatment systems are installed by government and donor agencies, much less data are available to analyse factors related to sustainability (or otherwise) of mostly demand-driven PoU water treatment systems. Reduced performance of the PoU water treatment system affects consumers and does not serve the purpose for which it is installed. The performance of any water treatment system is directly related to proper operation, periodic maintenance, and improved monitoring, and the membrane process is no exception.

Water is treated at the household's entry using PoE and PoU water treatment systems. PoE water treatment systems are deployed at individual consumer sites instead of purifying water before entering the reticulation (distribution/network of pipes) system. As a result, each family or small communal unit will require one PoE water treatment system. Compared to centralised treatment plants, the lower capital cost is the main benefit of PoE water treatment systems. Because they are installed on-site, the treated water does not sit in the reticulation system for a long time, reducing contamination risk. However, the fundamental issue with PoE is that constant monitoring of water quality is not cost-effective.

Rather than processing all incoming water to a household, the PoU water treatment systems are utilised to treat water at a single tap. These water treatment systems handle a limited amount of water and are typically installed beneath the kitchen sink, supplying only that tap with treated water. While a centralised water treatment plant is operated and managed by a dedicated organisation responsible for the operation, maintenance and performance monitoring, the PoU water treatment system needs to be managed and operated by consumers. Consumers normally are not aware of the operation and maintenance (O&M) of centralised water treatment plants; however, they need to be aware of the operation and periodic requirements of PoU water treatment systems maintenance. Due to the availability of outsourcing agencies (usually agencies supplying and installing the PoU water treatment systems), periodic maintenance of these units is much easier. PoU water treatment systems can have operational failure and functional failures. Operational failures affect water flow from the systems; hence such failures can be detected immediately. In case of functional failures which cannot be detected early and easily, the system continues to operate, but contaminants are not effectively removed.

PoU water treatment systems have emerged as a complementary alternative to centralised water treatment plants for households, commercial establishments, educational institutions and small communities. The maintenance of PoU water treatment systems is not adequately defined and regulated. Technical and commercial management of PoU water treatment systems can be categorised as follows:

- planning and development
- design and fabrication
- installation
- operation
- maintenance and performance monitoring

The planning, development, design, and installation mechanism is relatively well established. The commercialisation of PoU water treatment systems has brought market-driven factors that contribute to improved consumer awareness about the operation.

Commercially available membrane-based PoU water treatment systems include a wide range of options. Consumers frequently install these water treatment systems and then use them for years without ever checking the water quality. However, most PoU water treatment systems have a limited lifespan, making them unreliable after extended use. To ensure that they work properly and that the water is safe to drink, they must be maintained regularly. Furthermore, because PoU water treatment systems are exclusively installed in the kitchen, consumers should be aware that they depend on PoU water treatment systems to meet their drinking and cooking water demands. They cannot drink the water from any other sources without compromising their health. There is a rational change in the use of PoU water treatment systems in developing countries, and people who can afford them are demanding systems which can remove almost all the contaminants. Considering such requirements and improved awareness about water contamination and adverse health effects, demand for PoU water treatment systems, particularly those based on membrane processes, is steadily rising. Moreover, many variants are being increasingly offered by manufacturers and vendors. Some developing countries consider membrane-based PoU water treatment systems to be synonymous with PoU water treatment systems. The main concerns about PoU water treatment systems are whether they will be able to provide safe water continuously and health concerns associated with failing to purify all incoming water to a home.

The efficient O&M of membrane-based PoU water treatment plants is essential for all these requirements.

6.2 OPERATION AND MAINTENANCE RELATED CHALLENGES OF POINT-OF-USE WATER TREATMENT SYSTEMS

O&M related aspects of these systems have also considerably improved as consumers demand continuous operation and willingness to pay. Water supplied to households may contain compounds such as suspended solids, microorganisms, organics, and minerals, particularly in developing countries. Regular maintenance of any membrane-based PoU water treatment system is essential to prolonging the system's life and ensuring the best performance. The filters (pre- and post-treatment) in the membrane-based PoU water treatment system must be replaced regularly. The period between filter replacements is determined by the volume, quality, and concentration of contaminants in the water to be treated. The contaminant concentration, membrane rejection percentages, and removal efficiency influence filter replacement. Replacement intervals can be determined with the help of manufacturers and dealers. Common issues requiring periodic maintenance are discussed in subsequent subsections.

6.2.1 Clogging of the pre-treatment unit

Water first passes through a pre-treatment unit to preserve the sensitive membranes (component). This pre-treatment unit (normally in the form of a cartridge in PoU water treatment systems) is designed to filter out suspended solids. Pre-treatment of the feedwater to membrane-based PoU water treatment system can minimise fouling of the membrane and thereby increase the system's overall recovery rate. The feedwater may contain suspended and dissolved solids. Suspended particulates can settle on membrane surfaces, obstructing feedwater channels and increasing system friction losses. Scaling can occur when dissolved materials precipitate out of the water. Pre-treated water entering the RO system might impair the RO pump's efficiency, resulting in higher energy consumption. Another purpose of pre-treatment is to reduce chlorine concentration in feedwater which otherwise can spoil the membrane (McMordie *et al.*, 2013).

Activated carbon and its derivatives are mainly used for pre-treatment. GAC and carbon block filters are commonly used for pre-treatment in membrane-based PoU water treatment systems.

The main operational challenge with water treatment is granular medium clogging, which results from the accumulation of various types of materials that reduce the media's infiltration capacity. Contaminated solids, microbial activity etc. influence total solids accumulation in granular media. Granular media clogging is a natural and progressive process; therefore, some degree of clogging is unavoidable. When a membrane-based PoU water treatment system receives highly turbid water, pre-treatment units become overloaded with suspended solids. They quickly clog and require frequent washing. The finer the filter, the more particles are trapped. Thus, the filter must be replaced more frequently. The filter can easily clog if the pore size of the filter media is too small or the density of suspended solids in the untreated water is too high, and it will need to be replaced frequently. On the other hand, suspended solids may flow through the water treatment system if the pore size is too large. This can decrease output and degrade the quality of treated water. However, the issue emerges when advanced blockage limits treatment efficiency and severely shortens the system's lifespan.

6.2.2 Clogging/fouling of membranes

The membranes of the PoU water treatment system can clog by several factors described in subsequent sub-sections. Water flow in the system might reduce due to clogging of the membranes, resulting in poor system performance.

Fouling is defined as the accumulation or adhesion of retained particles, colloidal particles, macromolecules, salts, and other contaminants on the surface of the membrane and/or agglomeration in the pores, resulting in partial or entire obstruction of the pores and a steady drop in flow (Figure 6.1). The different ways the pores get blocked are determined by the size and type of the solute in relation to the membrane's pore-size distribution. The likelihood of creating a deposit on the membrane surface increases permeation resistance when single macromolecules or groups partially seal holes. Internal pore blockage occurs when substances are accumulated or adsorbed on the membrane pores' interiors, decreasing the space accessible for permeate flow. When particles (larger than the membrane pores) accumulate over the surface of the membrane and obstruct them, complete blockage of the pores occurs.

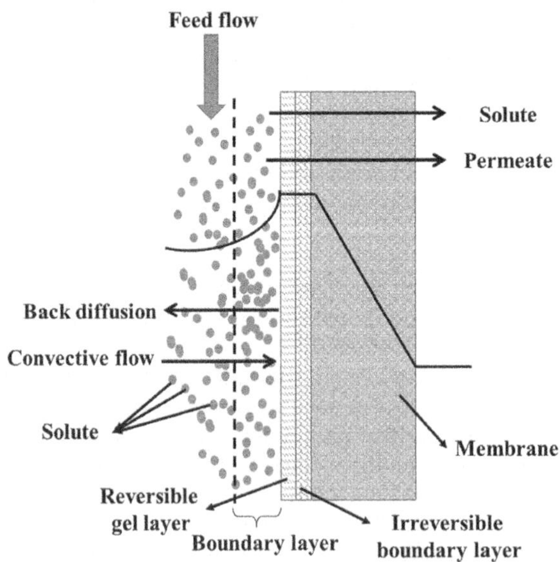

Figure 6.1 Membrane fouling.

Flux, recovery, pre-treatment, chemical cleaning, and, in the case of low-pressure membranes, hydraulic backwashing are all operational factors that influence membrane fouling.

Fouling, unlike concentration polarisation, which is a reversible phenomenon independent of the operational time of the membrane-based systems, is an irreversible phenomenon that is time-dependent. Fouling occurs when contaminants collect on the surface or in the pores of a filter membrane (Pandey *et al.*, 2012). Foulants (disruptive chemicals) block water passage across the membrane, increasing hydraulic resistance and energy consumption and possibly causing damage to the membrane and other system components. Major types of membrane foulants include:

- Particulate and colloidal fouling
- Scaling/inorganic/precipitation fouling
- Organic or biological fouling

Fouling can be reversible, such as when a simple washing can expel the foulants and restore full function to a membrane, or irreversible, such as when foulants chemically bond with the membrane material and permanently degrade its performance. Fouling has been described as a reduction in the active area of the membrane, resulting in a reduction in flux below the membrane's theoretical capacity for the given driving power. This is true if the pores are obstructed or blocked, but a cake layer on a membrane's surface is a resistance in series with the membrane resistance. Two types of chemicals cause problems: those that damage the membrane and pollute it. Because a fouled membrane must be cleaned, harm to the membrane may occur if caution is not exercised during the cleaning process (Yu *et al.*, 2017). Fouling during filtering has a clear detrimental impact on the economics of any membrane process; thus, it must be recognised, and precautions are taken to limit the consequences. Fouling can be severe in MF and UF, with the process flow often being less than 5% of the pure water flux (Field, 2010).

6.2.2.1 *Types of fouling*
6.2.2.1.1 Particulate and colloidal fouling
Colloidal fouling refers to particles of diameters ranging from 10 nm to 1 μm (Ismail *et al.*, 2018). Large proteins and viruses are frequently distinguished on the smaller end of the spectrum. Inorganic colloids include silicate minerals (including clays), iron and aluminium oxide particles, and silt. Humic material aggregates, cell fragments, and biopolymers, essential components of extracellular and intracellular polymeric substances such as polysaccharides (carbohydrates) and proteins, make up organic colloids. They are both capable of clogging membranes.

The rate of coagulation is mostly determined by colloidal stability. The colloidal stability is influenced by two key factors: salt content and pH. Stronger colloidal stability represents the surface charge of the colloidal components in the feed water. Ion-exchange softening and pH modification are two methods for increasing colloidal stability. When the stable colloidal's crossflow velocity reduces momentarily through the membrane (mainly UF), the starting flux is being recovered with an increased flow velocity again; however, the flux is not recovered while processing an unstable colloidal dispersion. When the trans-membrane pressure is momentarily elevated, the reaction is comparable. When the pressure is reduced, the stable feed water flux returns to its original value, whereas the flux of an unstable colloid solution decreases. When treating unstable colloidal dispersions, extreme caution must be exercised to ensure that the operating circumstances remain constant, whereas stable solutions' reversible flux behaviour allows for changes in the operating conditions.

6.2.2.1.2 Scaling/inorganic/precipitation fouling

The accumulation of crystalline salts, oxides, and hydroxides in the feed water causes scaling, also known as inorganic or precipitation fouling. Membrane scaling arises when soluble particles precipitate from the water and accumulate on the membrane's surface or clog in its pores. The accumulation of particles on a membrane that causes it to clog is called scaling. Because the membranes need to be cleaned more frequently due to scaling, they would use more energy and have a shorter life expectancy. The nominal flow is reduced as a result of scaling. The disadvantages include increased energy consumption, increased cleaning frequency, and a shorter membrane life span. When a solution becomes increasingly concentrated against the feed side of the membrane and eventually exceeds the feed water saturation threshold, ionic constituents fall out of the feed water and crystallise and/or bind to the surface of the membrane, causing precipitation fouling. Scaling is a concern for RO/NF systems with high conversion rates, particularly when the feed stream contains high calcium or magnesium concentrations.

6.2.2.1.3 Organic fouling

Biofouling or organic fouling (abiotic), whereas fouling induced by organic matter formed from microbial cellular detritus is abiotic biofouling. Figure 6.2 depicts the steps involved in biofouling. Biofouling is the membrane process's 'Achilles heel' because the microbes can grow over time even after removing 99.9% from feedwater. After the removal of 99.99%, there might be sufficient cells available to grow the microbes at the expense of biodegradable substances available in the feed water (Nguyen *et al.*, 2012). When bacteria attach to the membrane, they proliferate and digest the nutrients provided, eventually

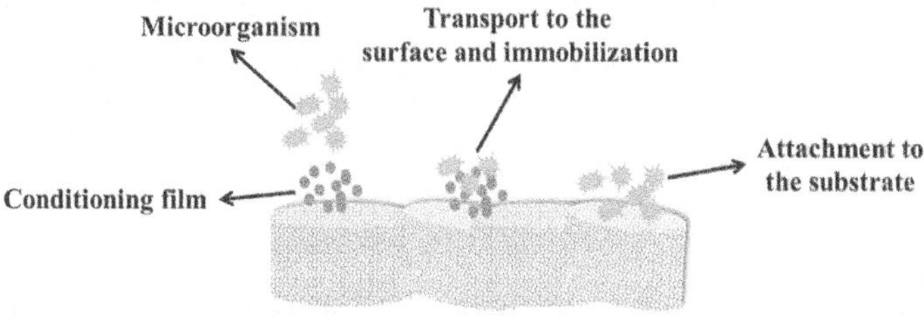

Figure 6.2 Steps in the organic fouling.

producing a biofilm when cell density reaches a crucial level. Biofouling has been identified as a key contributor to more than 45% of all membrane fouling and has been described as a significant issue in NF and RO membrane filtering.

Biofouling degrades the efficiency of membranes and necessitates costly cleaning techniques to eliminate biofilms. The impact of biofilms on plant performance is related to the biofilm's structure and content. Microorganisms, particularly bacteria, are the primary cause of biofouling, and because bacteria are extremely versatile, they can colonise practically any surface under extreme conditions such as temperatures ranging from −12°C to 110°C and pH values ranging from 0.5 to 13 (Maddah and Chogle, 2017).

6.2.3 Failure of post-treatment

Post-treatment is common in RO membrane-based PoU water treatment systems. As explained earlier, the RO membrane can remove almost all the constituents from water. A minimum level of dissolved solids (mineral content) can be achieved in treated water by (a) bypassing part of feed (raw) water after disinfection (the most preferred option is UV disinfection) and adding to permeate (b) by addition of chemicals.

UV lamps and other components such as quartz glass tubes can get damaged for several reasons affecting the efficiency and function of the UV disinfection unit. If the second option of maintaining dissolved solids is exercise, chemicals continuously added to permeate can get exhausted.

6.2.4 Non-functional sensors

Various sensors are being installed in PoU water treatment systems to indicate the system's functioning. LED indicators are normally used in these systems. In addition, flow rate indicators are also provided, and any reduction in flow rate is also indicated in the system. Reduction in flow due to several reasons, such as clogging of GAC/carbon filters, is also indicated using these sensors. These sensors help even in recognising the need for maintenance of the system.

Failure of PoU water treatment system sensors for any reason is also a possibility. Failure of these sensors/indicators can adversely affect the system's functioning. This may result in possible skipping of maintenance of any component of PoU water treatment systems, which otherwise could have been done had the sensors been working.

6.3 PREVENTIVE MAINTENANCE OF MEMBRANE-BASED POINT-OF-USE WATER TREATMENT SYSTEMS

PoU water treatment systems come with various treatment options, including filtration, disinfection, and desalination with RO membranes. As a rule of thumb, membrane-based PoU water treatment systems normally have only one stage. A generic but elaborate schematic of a membrane-based PoU water treatment system is shown in Figure 6.3. Without going into depth about each stage of the RO process (because the number of stages varies by model), basic maintenance information that applies to most membrane-based PoU water treatment systems is discussed in subsequent subsections.

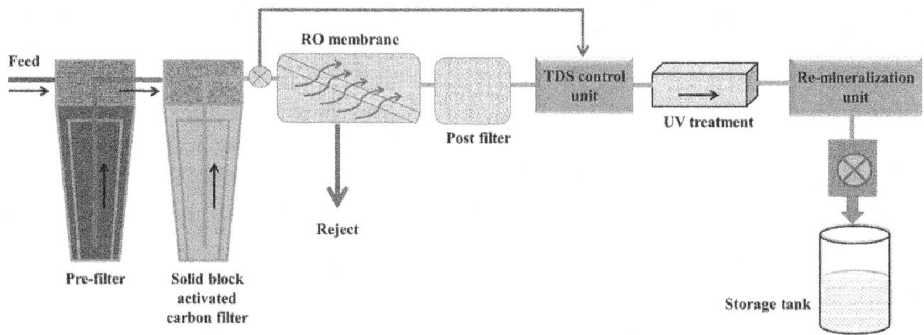

Figure 6.3 Membrane-based PoU water treatment system.

Processes detailed in this section apply to membrane-based water treatment plants (including large plants) and may not be used in situ (at the PoU water treatment system). However, maintenance processes detailed in subsequent sections should generally be used either at the site of PoU water treatment, or individual units can be taken to centralised facilities for inspection/cleaning. Due to longevity and reduction in the cost of PoU water treatment systems components, replacement is now preferred over repair/cleaning as this may take a longer period, and the component's performance is also likely to get affected.

Preventive maintenance is a prerequisite for improving PoU water treatment system performance like any other system. Because these systems have matured over the years, operation failures are fairly well-known. Sensors and indicators have also improved these systems' functioning and increased life. Moreover, the failure of these systems is often predictable, and preventive maintenance is fairly routine. An emergency can be prevented with periodic maintenance of membrane-based PoU water treatment systems. We can identify problems with regular inspections and avoid stoppage of the system, water quality concerns, damage to the components, or costly damage. Consequently, repairs of the systems could be the replacement of components such as membranes or pre-treatment filters. Hence, preventive maintenance has become essential in properly operating and maintaining membrane-based PoU water treatment systems.

Installation of the PoU water treatment system should be properly undertaken to avoid any possible O&M problems later. Most PoU water treatment systems have detailed instructions for installing the system, possibly the most important step in the entire process. Plumbing to a water source is also an important installation step and needs to be carefully executed. There is no significant skill required to operate a membrane-based PoU water treatment system. Operators' manual is available with most of the systems, and vendors normally explain the operation of these systems during the system installation.

Following components are provided for preventive maintenance of PoU-based water treatment systems by some manufacturers.

- **Flushing system**: Flushing at regular intervals keeps the surface of the membrane clean and offers longer life. So, automatic flushing systems are included in the system.
- **Tank level control**: After the tank is filled, the system needs to be shut off. The solenoid valve is used to control the level in the tank. When the tank is full, this gives the signal to the booster pump to stop.
- **Microcontroller**: All automatic functions like flushing of RO membrane at regular intervals. Tank level control is controlled by a microcontroller.
- **SMPS power supply** – It gives constant voltage output even at variable input AC voltage and voltage fluctuation. All water purifiers select SMPS power supply.

6.3.1 Maintenance of pre-treatment unit

Since membranes are sensitive to fouling, pre-treatment of the feed is necessary to prevent colloidal, chemical, and biological fouling and scaling. The treatment plan must control membrane fouling to the point where a reasonable cleaning frequency may be attained. A variety of pre-treatment procedures are now employed for low-pressure membranes.

It is observed that pre-treatment units consisting mainly of cartridges and PAC filter often gets chocked and need to be physically cleaned using water. GAC filters can be washed, and the flow rate can be improved by periodical cleaning with treated water. Carbon block filters are currently replaced after exhaustion as the manufacturers supply these filter blocks separately. Every 6–9 months, this pre-treatment unit should be replaced. The pre-treatment unit can foul or become clogged if not properly maintained or changed regularly, rendering it unable to protect the RO membranes. These systems are compact and normally sealed, making them difficult to open, and replacement is the only option.

6.3.2 Maintenance of membrane unit

Maintenance of membrane-based PoU water treatment system can be accomplished by pre-treating the feed to reduce its fouling potential, enhancing the antifouling characteristics of the membranes, increasing membrane cleaning and backwashing conditions, and optimising operating parameters.

There are three signals which indicate that membrane-based (particularly RO) water treatment systems require maintenance, viz. low water pressure, odour, colour or taste in water and continuous constant flow at the outlet of the PoU water treatment system. These signals can easily be noticed if sensors or indicators are not provided on PoU water treatment systems.

- Low water pressure can easily be realised if the volume of water dispensed through the system is low and pressure is considerably reduced. This can happen due to multiple reasons, and clogging of pre-treatment is the major reason. Blockage of membrane might also be another reason for low pressure.
- Treated water from membrane-based PoU water treatment system has typically no perceivable odour, taste or colour and the presence of any

of these attributes often indicates the necessity of maintenance of the systems. If treated water smells of chlorine, the pre-treatment unit or membrane requires replacement. Chlorine can damage the membrane instantaneously.
- Continuous constant flow at the PoU water treatment system outlet may indicate physical damage to the membrane, and water is flowing out almost at the same flow rate. It also means that the pre-filter is not performing its function and may require immediate replacement.

The membrane-based PoU water treatment system can be disinfected regularly with the manufacturer's recommended products. It is feasible to clean and renew the membrane if membrane fouling is recognised early; the approach depends on the kind of membrane and fouling. Membranes that are completely blocked or torn must be replaced. Damaged RO membranes, on the other hand, are difficult to identify. Treated water can be tested periodically to see if the membrane is still intact and working properly. Many systems have a monitor/indicator that displays a high TDS content or insufficient TDS rejection, which is one sign of improper operation. Pre-treatment of the water using a softener can extend the membrane's life in cases when the water is relatively harsh.

Many systems are designed with an auto-flush restrictor that automatically flushes the membrane-based (RO) water treatment system for 30–45 s at the start and once every hour/2 h when the system is in operation.

6.3.2.1 *Membrane unit cleaning*
Flushing or chemical techniques such as improved backwashing can remove fouling compounds. Cleaning operations are defined as either cleaning in place (CIP) or chemical cleaning off-line (or soaking). CIP cleaning involves the membrane module cleaning without removing it from the installation, whereas off-line cleaning involves removing the module from the system and immersing it in a chemical.

Membrane cleaning entails transferring large amounts of chemicals to the fouling layer and returning the reaction products to the bulk liquid phase. As a result, hydrodynamic conditions must be created during cleaning that encourages contact between cleaning agents and fouling compounds. Dynamic cleaning, which involves circulating cleaning solutions throughout the system, can be more effective than static cleanings, such as soaking, in mass transfer. The frequency of cleaning the membranes has been projected from process optimisation. A cleaning process can be optimised in order for:

- Enhancing or sustaining cleaning efficiency,
- Minimising chemical, water, and energy usage, and
- Reducing the waste effluent impact on the environment. Typically, optimisation is investigated by modifying one chemical or physical condition at a time.

Cleaning, however, involves several parameters due to the complicated nature of the contamination, including process sequence, temperature and pH of the solution, the quantity of the chemicals, hydrodynamic conditions and cleaning duration, among others. According to the one-parameter-at-a-time strategy, a thorough understanding of a membrane application requires many trials to cover all parameters.

However, membrane cleaning is often carried out ex-situ by taking spent membranes to centralised facilities instead of carrying them out in the household.

6.3.2.1.1 Choosing type of cleaning

The requirements for various cleaning procedures are based on the process. The order in which these stages are completed impacts the overall cleaning effectiveness. Backwashing followed by forward flushing, for example, is more successful than the other way around. Backwash could dislodge otherwise difficult-to-reach foulants in the pores, with forward wash then helping in their removal from the stream. To efficiently remove macromolecular and mineral foulants, an alkaline cleaning followed by an acid stage is often used in the dairy and surface water sectors, respectively. Forward flush, backward flush, and air flush are three distinct membrane cleaning methods (www.lenntech. com/membrane-cleaning.htm).

6.3.2.1.1.1 Forward flush

Feed water or permeate is used for flushing the membranes forward when using forward flush. The water or permeate flows through the system at a faster rate during the feed water or permeate flow phase than during the production phase. As a result of the quicker flow and turbulence, particles that the membrane has absorbed are freed and ejected. Particles absorbed through membrane perforations are not allowed to escape. The only method to get rid of these particles is to flush them backwards.

When the forward flush is utilised, the barrier responsible for dead-end management in a membrane is opened. The membrane performs crossflow filtering for a brief period without creating the permeate. The purpose of a forward flush is to generate turbulence to remove a built-up layer of contaminants from the membrane. A substantial hydraulic pressure gradient is necessary for forward flushing.

6.3.2.1.1.2 Backward flush

The permeate is pushed into the feed water side of the system under pressure, providing double the flux used during filtering. A chemical cleaning approach might be applied if the flux has not completely repaired itself after back flushing. When a backward flush is utilised, the membrane's pores are flushed from the inside out. The pores are cleansed because the pressure on the permeate side of the membrane is larger than the pressure inside the membranes.

A backward flush is carried out at a pressure around 2.5 times higher than the output pressure. Permeate is always used for a backward flush because the permeate chamber must always be free of pollution. Backward flushing causes a reduction in process recovery. As a result, a backward flush must be completed in the quickest time feasible. However, the flush must be maintained going long enough to flush a module's volume at least once. A backward flush is a filtration procedure that reverses the direction of filtration.

6.3.2.1.1.3 Air flush or air/water flush

The so-called air flush or air/water flush is a modern cleaning method. This is a forward flush that involves injecting air into the supply line. The use of air (while the water speed remains constant) creates a significantly more turbulent cleaning method. Fouling on the membrane surface must be eliminated as thoroughly as possible during reverse flush. The air flush, which Nuon created in collaboration with design hourly volume and crossflow, has proven to be quite effective in this process.

Air flushing is flushing the inside membranes with a mixture of air and water. When you add air to the forward flush during an air flush, air bubbles form, which increases turbulence. As a result of the turbulence, fouling is removed from the membrane surface. Compared to the forward flush, the air flush uses less pumping capacity throughout the cleaning process.

6.3.2.1.1.4 Chemical cleaning

During a chemical cleaning process, membranes are immersed in a solution of chlorine bleach, hydrochloric acid, or hydrogen peroxide. The solution soaks for a few minutes in the membranes before being flushed out with a forward or backward flush to remove the contaminants.

Chemical cleaning of the membranes is required when the earlier methods fail to reduce the flow to an acceptable level. Chemicals like hydrogen chloride (HCl) and nitric acid (HNO_3), as well as disinfectants such hydrogen peroxide (H_2O_2), are added to the permeate during reverse flushing. Once the entire module has been filled with permeate, the chemicals must soak in. After the cleaning chemicals have thoroughly soaked in, the module is flushed and eventually put back into production.

Cleaning techniques are frequently mixed. For example, pore fouling can be removed with a backward flush, followed by a forward or air flush. Many factors influence the cleaning process or strategy employed. The most appropriate methods are discovered by trial and error (practice tests).

6.3.2.1.2 Parameters for cleaning operation

Cleaning is frequently only partially effective, particularly when flow channels are obstructed and huge portions of the elements are inaccessible to the recirculating cleaning solution. As a result, the use of innovative antiscalants and antifoulants and sufficient pre-treatment and pilot testing of cleaning processes created during the pilot testing stage should reduce or eliminate the need for cleaning. When cleaning is required during operation, it should be done as soon as possible after fouling has occurred.

Membrane manufacturers and practitioners generally agree that RO systems should be cleaned before the following performance changes occur:

- A 10–15% reduction in normalised permeate flowrate.
- A 10–15% increase in differential pressure.
- A 1–2% reduction in salt rejection.

If a cleaning technique fails to fully restore system performance to the reference RO system startup values, the same cleaning procedure will certainly result in a faster reduction in system performance and a greater cleaning frequency in the future. As a result, it is critical to address two issues at this time:

- Develop a better cleaning process, and
- Look into possible pre-treatment improvements to avoid membrane fouling.

Cleaning and process improvement activities should be continued until the RO performance is stable. Even with well-piloted and constructed RO plants that run well at first, the source water quality will invariably vary with time. Changes in equipment and personnel impact performance, necessitating constant vigilance and readiness for the plant's continuous improvement.

6.3.2.1.2.1 Choosing cleaners

Membrane manufacturers currently propose certain sorts of cleaning agents. Some are commercial chemicals, while others are proprietary cleaners. The membrane must be cleaned chemically to regain the majority of its permeability. When flushing and/or backwashing fail to restore the permeate flux, chemical cleaning is used. Chemical cleaning normally has a larger chemical dose than enhanced backwashing, and chemical cleaning is usually done less frequently (approximately once a week). Furthermore, enhanced backwashing can be automated due to its off-line operation, whereas chemical cleaning requires manual effort. It is critical to choose the right chemical cleaning agents, use them under the right conditions, and understand how they work. Cleaning agents are often chosen based on the sorts of foulants present.

Various generic chemicals are advised for blending at the site where major membrane manufacturers make cleaning solutions for five categories of foulants: acid-soluble foulants, bio-film/bacterial slime/biological matter, carbon-containing oils/organic matter, dual organic and inorganic coagulated colloids, and silica and silicates are the five categories of foulants.

Commercially available booster cleaners are offered to improve the efficiency of the generic cleaners prepared on site. Chemical companies that specialise in RO operations offer a wide range of unique RO membrane cleaners for convenience and technical assistance. When generic cleaners fail to meet expectations, proprietary cleansers and cleaning support are available.

6.3.2.1.3 Cleaning procedure

There are six steps in the cleaning of membrane elements in place in RO (membrane)-based water treatment systems:

1 **Mix cleaning solution.**
2 **Low flow pumping**
   To displace the process water, pump preheated cleaning solution to the vessels at a low flow rate (approximately half of that given in Table 6.1) and low pressure. Use only enough pressure to compensate for the pressure drop from feed to concentrate with the RO concentrate throttling valve entirely open to decrease pressure during cleaning. The

Table 6.1 Recommended re-circulation flow rates during cleaning.

Feed Pressure* (kPa)	Element Diameter (m)	Feed Flow Rate per Vessel (m^{-3} h^{-1})
138–414	0.0508 (2 in)	0.68–1.14
	0.1016 (4 in)	1.82–2.72
	0.1524 (6 in)	3.64–4.54
	0.2032 (8 in)	6.81–9.09

Source: Ning (2012).
*Dependent on the number of elements in the pressure vessel.

pressure should be low enough to prevent the formation of permeate. Low pressure reduces dirt re-deposition on the membrane.

3 **Recirculate**

A cleaning solution will be present in the concentrate stream after the process water has been displaced. Allow the temperature to settle before recirculating the concentrate to the cleaning solution tank.

4 **Soak**

Turn off the pump and let the elements soak. A soaking period of roughly 1 h is sometimes sufficient. An overnight soaking duration of 10–15 h is beneficial for foulants that are tough to clean (Ning, 2012). Use a slow re-circulation rate to maintain a high temperature for an extended soaking period (about 10% of that shown in Table 6.1).

5 **High flow pumping**

Feed the cleaning solution for 30–60 min at the rates listed in Table 6.1. The high crossflow rate flushes out the foulants removed from the membrane surface by cleaning with limited or no permeation through the membrane to avoid foulant compacting. If the elements are significantly clogged (which should not be the case), a flow rate 50% greater than given in Table 6.2 may help with cleaning. Excessive pressure drop may be an issue at greater flow rates. The maximum pressure drop used is 138 kPa per element or 414 kPa per multi-element vessel, whichever is more restricting.

6 **Flush out**

Unless there are corrosion issues, such as seawater corroding stainless-steel pipelines, pre-filtered raw water can be used to flush away the cleaning solution. The minimum flush temperature is 20°C to avoid precipitation.

6.3.2.2 Anti-foulant chemical design

Membranes that have become fouled require physiochemical treatment, reducing membrane lifespan and increasing overall operational costs. Scaling, particle fouling, and microbiological fouling – the three types of fouling mechanisms outlined above – are all covered by the term antifoulant in its broadest sense.

Table 6.2 Recommended troubleshooting steps for membrane-based PoU water treatment systems (www.purewaterproducts.com).

Symptoms	Possible Causes	Corrective Action
Low inlet pressure	Low supply pressure	Increase inlet pressure
	Cartridge filters plugged	Change filters
	Solenoid valve malfunction	Replace solenoid valve and/or coil
	Motor may not be drawing correct current	Use clamp-on amp meter to check the motor amp draw.
	Concentrate valve might be damaged	Replace needle valve
	Leaks	Fix any visible leaks
Low permeate flow	Low inlet flow	Adjust concentrate valve
	Cold feed water	See temperature correction sheet
	Low operating pressure	See low inlet pressure
	Defective membrane brine seal	Inspect and/or replace brine seal
	Fouled or scaled membrane	Clean membranes
High permeate flow	Damaged product tube o-rings	Inspect and/or replace
	Damaged or oxidised membrane	Replace membrane
	Exceeding maximum feed water temperature	See temperature correction sheet
Poor permeate quality	Low operating pressure	See low inlet pressure
	Damage product tube o-rings	Inspect and/or replace
	Damaged or oxidised membrane	Replace membrane
Membrane fouling	Metal oxide fouling	Improve pre-treatment to remove metals. Clean with acid cleaners.
	Colloidal fouling	Optimise pre-treatment for colloid removal. Clean with high pH anionic cleaners.
	Scaling ($CaSO_4$, $CaSO_3$, $BaSO_4$, SiO_2)	Increase acid addition and antiscalant dosage. Reduce recovery. Clean with acid cleaners.
	Biological fouling	Shock dosage of sodium bi-sulphate. Continuous feed of sodium bi-sulphate at reduced pH. Chlorination and de-chlorination. Replace cartridge filters.
	Organic fouling	Activated carbon or other pre-treatment. Clean with high pH cleaner.
	Chlorine oxidation	Check chlorine feed equipment and de-chlorination system.
	Abrasion of the membrane by crystalline material	Improve pre-treatment. Check all filters for media leakage.

The Langelier saturation index (LSI) is a water balance measurement based on calcium carbonate saturation. It evaluates if the water is corrosive (low LSI), balanced, or scale-forming (high LSI) (Arnal *et al.*, 2011).

6.3.2.2.1 Scale control

Antiscalants have been developed and employed in boiling water and cooling water chemistry and used in boilers, evaporators, cooling towers, and cooling systems. In RO membrane-based water treatment systems, anionic polymers, polyphosphates, and organophosphorus chemicals, also known as threshold inhibitors and dispersants, are utilised in sub-stoichiometric levels, usually in the range of 1–5 mg L^{-1} concentrations. The crystallisation rates from supersaturated solutions are slowed, and crystal-packing orders are changed by binding to the surfaces of developing crystal nuclei. Although super-saturation of solutes in the water will eventually equilibrate through crystallisation, there will be little or no scale formation throughout the residence duration of the water in the system due to this mechanism. The residence duration is very brief (a few seconds), the concentration of seed crystals is minimal, and the temperature is consistent, making RO stand out among water conditioning systems. As a result, higher levels of super-saturation can be achieved without crystallisation. Limits of saturation and scaling rates, on the other hand, are difficult to model, quantify, and anticipate. Other organic or inorganic solutes in the water cause interference.

6.3.2.2.2 Controlling colloidal fouling

The work is much more difficult because of the range of potential foulants and the complexity of their interactions in the same water and with the membrane. The stability and agglomeration of colloidal particles are critical in both natural and industrial fluids. Antifoulant development is progressing gradually, based on basic colloidal science and testing of model foulants proposed by RO foulant analysis data.

6.3.2.2.3 Controlling bio-fouling

The literature on water purification systems is substantial. Much of the art and science discovered to be effective can also be applied to PoU water treatment systems. There are some factors unique to the RO system that should be mentioned. The thin, salt-rejecting polyamide or cellulose acetate barrier membrane must be chemically compatible with the chemicals used to sterilise and clean the system. The accumulation and exponential growth of microorganisms within the system should be avoided as much as possible. Pre-treatment of feedwater, proper upstream unit maintenance, continuous flow of water through the RO unit, a thorough monitoring and sanitisation programme, and the use of preservatives during downtime are all critical to this goal.

6.3.2.2.4 Antifouling coating of membranes

Coating the membrane with an antifouling material is one way to reduce membrane fouling. These coatings are designed to extend the period between membrane cleanings and improve flux recovery after cleanings. Antifouling coatings adhere to the 'Whitesides' guidelines', which include making the membrane surface more hydrophilic (but not too hydrophilic), adding hydrogen-bond acceptors, eliminating hydrogen-bond donors, and maintaining an overall neutral electrical charge. These coatings are highly effective because they generate a thick hydration layer that prevents foulants from sticking to the membrane surface. It is critical to control hydrophilicity to avoid excessive surface free energies that attract foulants (e.g., aiming for a 35–45° water contact angle).

Polyethene glycol (PEG)-based coatings are probably the most prevalent antifouling chemical. PEG-based coatings can be applied on membrane surfaces in a variety of methods. UV radiation or controlled radical grafting technologies such as atom transfer radical polymerisation can be used to graft PEG-based coatings to or from a membrane. Unreacted carboxylic acid/acyl chloride groups on the membrane surface can be linked with modified PEG. Physical adsorption, including dip-coating, can be used to apply PEG. On the other hand, PEG tends to deteriorate under standard membrane cleaning conditions, such as hypochlorite solution treatment.

6.3.3 Maintenance of post-treatment

Bypass is often observed as the only post-treatment practised for remineralisation, which does not require much maintenance as only a small proportion of water is bypassed. In case specific chemicals are used, these chemicals are dosed in water to maintain concentrations of these minerals/dissolved solids in treated water. These chemicals (in solid form) are periodically checked visually and added if exhaustion of these chemicals is visible.

An indicator/sensor is increasingly being used if a UV lamp is not properly functioning. Consumers can report if a UV lamp is indicated to be non-functional, and the replacement of a UV lamp remains the most viable option in most cases. Major manufacturers have devised their O&M strategy and trained crew for maintenance personnel. This personnel are provided training and sometimes training for periodic maintenance and repair. Maintenance and training are fairly standardised, and component replacement of the PoU water treatment system remains the major solution. Continuous disinfection systems either address an existing bacterial problem or protect against bacterial contamination in the future. Where the water enters the house, a UV steriliser or UV light is installed (PoE). UV light is more effective when the water is clear. Pre-treatment with water filtering is required to remove particles that could prevent UV radiation from killing germs. The lamp has a quartz sleeve that must be maintained clean, and the light must be replaced once a year.

As chemical dosing is carried out automatically, exhaustion of these chemicals is automatically indicated. This indication helps in refilling chemicals. In the second arrangement, a small proportion of raw water after disinfection (UV treatment) is added to the treated water. As no moving part is involved in this arrangement, maintenance is limited to the life of the UV disinfection system. Two ways to carry out remineralisation and chemical consumption depend on the quantity of these chemicals added to treated water.

A carbon filter stage is commonly added to some membrane-based PoU water treatment systems to 'polish' the water at the end of the cycle. This stage removes any lingering flavours or odours from the water. Every 6–9 months, these carbon filters should be replaced.

6.4 TROUBLESHOOTING OF POINT-OF-USE WATER TREATMENT SYSTEM

Troubleshooting means a systematic approach to problem-solving and is often used to identify and rectify operational issues with any system. With the advent of software, sensor, indicator, and IoT enabled systems, troubleshooting PoU water treatment systems is much easier. Moreover, the cost of maintenance by a specialist mechanic by visiting a household is also very high, which prompted the development of an innovative or easy-to-maintain system. Instructions provided by manufacturers are easy to follow for troubleshooting any system. A visit to maintenance crew is only warranted if the fault/issue persists in the system, and it is not easy to troubleshoot the system even after scrupulously following the instructions. Recommended troubleshooting steps for these systems are presented in Table 6.2 (www.purewaterproducts.com).

Preventive maintenance of PoU water treatment systems can be carried out by following the steps as narrated below.

6.4.1 Consistently replace reverse osmosis pre-filters

When a softening agent filters the RO intake, the softener's regeneration needs to be properly maintained. However, that can be hard if the consumer does not understand how either function. During the regeneration cycle of water softeners, if soft water is not produced, it implies that hard water minerals can pass through. Hence, the consumer must make sure the system is scheduled to regenerate during low usage hours. Generally, sediment and carbon pre-filters should be changed every 6–12 months to prolong the life of membranes.

6.4.2 Test hardness of feed water

To gauge how hard the RO unit is going to have to work to filter water, it is a good idea to test the hardness of the water feed. It should be monitored to check if RO maintenance needs to be revised due to changes in the water quality.

6.4.3 Test pressure of feed water

Since RO is a pressure-driven process, the membrane elements need a specific pump pressure to produce the proper product flow and rejection. Low inlet pressure results in a slow filling tank, more reject water and less quality drinking water. Regularly notating the pump pressure helps determine if a problem exists.

If the pressure is 300 kPa or less, the pressure should be compensated. Even though the RO process can operate fine on usual local water pressure (400 kPa), they perform even better when a small pump is used to boost the pressure to 400 kPa or higher.

6.4.4 Inspect feed water temperature

The temperature of incoming water also impacts the way the RO unit performs. The ratings of membrane elements are based on 25°C source water. Hence, it can be expected that the product flow to be reduced by 1.5% for every degree the feed water falls under this ideal temperature. Seasonal inconsistencies can be addressed with additional membrane elements or bigger products.

6.4.5 Test total dissolved solids of product water

Another important step of RO maintenance is comparing feed water and RO filtered water. When a water test shows similar hardness, there is something wrong. High levels of TDS imply that the produced water consists of harmful contaminants. Drinking it can pose health risks and even cripple cells' ability to absorb water molecules. In other words, low TDS levels make it easier for the body's cells to hydrate. If more than 80% of TDS are flowing through, the RO membrane needs to be replaced.

6.4.6 Test reject water

The RO process produces two types of water: purified water (or the product) and RO-reject (wastewater). When pressure is faulty, or the filtration system is not working adequately, it will reject more water than it purifies. This is why it is important to monitor reject levels and flow rates. Again, any changes should be investigated.

6.4.7 Replacement of membranes

The membranes can last up to 5 years when the water source is not too hard (or softening is present), and RO maintenance is a priority. The user can also play a factor. A water source with high ionic levels will wear on the membrane element more than a lower ionic water supply. Hence, the membranes should be changed on time.

6.4.8 Routine cleaning and disinfection

The purification process will always leave accumulated foulants on the feed side. Low pH (2–4) or high pH (7.5–11) cleaners can be used to remove deposits on the membrane that bacteria can adhere to. However, it is always best to call professional service. While a thorough clean will certainly aid and extend performance, it will not kill the present bacteria. This is why it is far more ideal for disinfecting the system completely. Doing so regularly decreases and controls microbial levels throughout.

6.5 SUMMARY

O&M related aspects of PoU water treatment systems have considerably improved due to the demand of consumers to operate these systems continuously. Water supplied to households contains suspended solids, microorganisms, organics, and dissolved solids. Periodic maintenance of any membrane-based PoU water treatment system is vital for prolonging the system's life and ensuring optimal performance.

With intense competition among PoU water treatment systems manufacturers, maintenance of these systems is fairly well developed, emphasising preventive maintenance and troubleshooting. The membrane remains the most important component of these systems, and membrane fouling should be avoided. Due to several operational and economic reasons, it is now preferred to replace rather than repair the membrane. Failure of the systems is also considerably reduced due to improvement in manufacturing processes and optimised operations.

doi: 10.2166/9781789062724_0151

Chapter 7

Techno-economic analysis of membrane-based point-of-use water treatment systems

7.1 INTRODUCTION

In most countries water is treated in centralised water treatment plants and transported and distributed to household's/collection points presuming that the distribution network remains intact and water of desired quality is delivered. The conventional water treatment plant is extensively used due to its significantly lower cost than other systems for centralised water treatment. Surface water remains the major water source for the conventional treatment plants for a community supply. Inorganic coagulants (such as aluminium sulphate and ferric chloride) are used in conventional plants and added to the raw surface water for the coagulation/flocculation process. Raw water passes through a cascade aerator before injecting the coagulant for natural aeration. The agglomerated flocks were introduced to the raw water and clarified or sedimented to separate the solids from the liquid. The final polishing is done with sand filters before the disinfection procedure begins. The disinfected water is then ready to be distributed to consumers as tap water. In the majority of the countries, disinfection is carried out using chlorination.

However, the reality is often different, particularly in developing countries, and water quality is compromised due to various reasons. Similarly, it is also presumed that the economy of scales favours centralised treatment and distribution of water in densely populated areas in cities. Contrarily, several case studies and examples are clear evidence that these water supply systems fail to achieve such ambitious objectives of supplying safe water mainly due to non-technical reasons. The water supplied by centralised water treatment plants has often shown measurable pathogens and other pollutants. In addition, water gets contaminated during distribution from centralised water treatment plants to households, particularly at PoU. We have provided several references, particularly from developing countries, in Chapter 2, clearly indicating deterioration in water quality from the water treatment plants to households. It is reported that water quality significantly deteriorates between source and stored water. The mean percentage of contaminated samples at the source was 46% as against 75% samples in household storage water. Piped water supplies show significantly less contamination than non-piped water, primarily due to residual chlorine. This indicates that despite efficiently treating water in centralised water treatment even from protected sources, water can get contaminated during distribution to households.

Moreover, improvement in water quality in centralised water treatment plants and subsequent transportation and distribution to households is an expensive proposition, considering that a very small fraction (only 3–5%) of household water is used for drinking and cooking. In addition, water supplied through a centralised system is used for commercial and, at times, industrial purposes making the actual use of water for domestic purposes even smaller. This brings forth an important question if water needs to be treated to the drinking water quality in a centralised water treatment plant or if other options such as PoU water treatment systems can be a better alternative to meet drinking water requirements. PoUs can be used to provide safe drinking water even from poor water quality sources.

If managed by community contribution or recovery of water tariff, the water supply system often fails in developing countries. In addition, contaminated water sources are not amenable to conventional water treatment of coagulation–flocculation, and filtration also requires different treatment solutions. Many water sources in developing countries are not sustainable and fail more than reported earlier, particularly due to inadequate protection against chemical or microbial pollutants. In the sparse community in a rural setting and informal settlements such as slums in urban and peri-urban areas in developing countries, PoU water treatment systems remain the only cost-effective solution to safeguard against water supplied from unprotected sources.

Engineering, economic and social factors like the cost of acquisition, operation, and maintenance of the treatment system or its energy requirement and social acceptance influence water treatment technology selection. In addition, water's chemical, microbiological, and physical characteristics also affect technology selection. There is a significant variation in the cost of water treatment from country to country and changes within a country, which largely depend on the following factors:

- water demand
- energy source and associated cost
- raw water quality
- country-specific/regional/local water quality standards
- preference for local versus imported treatment technology
- requirements (number) of water treatment units

The PoU water treatment system market combines multiple technologies mainly driven by commercial factors. Deployment of a simple PoU water treatment system is now restricted to a rural setting, mainly driven by donor funding. Membrane-based PoU systems are largely employed in an urban setting, and many combinations are emerging in the global market. Aesthetics is also given importance in addition to water treatment technologies to attract consumers.

Conventional water treatment systems are considered multi-barrier, whereas most household treatment technologies are single barriers. However, certain PoU water treatment systems, such as membrane-based, perform better than conventional water treatment systems for pathogen removal. The economy of scale has a bias towards some technologies such as membrane-based water treatment systems, for example RO, which has the flexibility to provide water from households to large communities. Some of the treatment technologies can only treat a specific contaminant, such as fluoride or arsenic, making it difficult to compare the cost of these technologies. As it is impossible to provide water treatment costs for each country and contaminants, a range of the cost of water treatment is provided.

7.2 COST-EFFECTIVENESS OF CENTRALISED WATER TREATMENT PLANTS

In developing countries, the prevalence of waterborne diseases is high. Waterborne pathogens are assumed to be present even in treated drinking water provided by the distribution system. In addition, centralised (conventional) water treatment plants with conventional and advanced treatment technologies appear to be an expensive solution.

According to a report published by CH2M Beca Limited (2010) for the Ministry of Health, New Zealand, the total capital cost for a medium-sized water treatment plant is 8 400 000 USD, with an annual operating cost of 180 000 USD (Moore *et al.*, 2017). The cost of UV treatment, pH correction, and chlorination at a water treatment plant is between 220 000 USD – 390 000 USD. A membrane UF plant would cost up to 200 000 USD. According to a 2007 survey, in Canada, centralised water treatment plants' operation and maintenance costs are 807 000 000 USD (Nelligan *et al.*, 2015). A PoU system costs between 67 and 318 USD to operate per year. It could cost a family ~1 200 USD to treat waterborne infections (Verhougstraete *et al.*, 2020).

The capital investment in rural water treatment plants has no realistic chance of being recovered. One of the major issues in guaranteeing the efficient operation and maintenance of big water treatment plants is a lack of technically skilled designers and operators. Although decentralised small-scale water treatment plants are thought to alleviate the problem of clean water supply, technical and economic challenges remain. The advent of such alternative technologies has been encouraged by the operational problems of conventional water treatment plants. Membrane-based systems have been the most popular choice for large-scale centralised water treatment plants. Polymeric membranes have become much less expensive due to economies of scale and improved manufacturing procedures. A big UF membrane system has a capital cost of around 6% higher than a traditional water treatment facility.

Compared to these costs, PoU water treatment systems are much more feasible since they are considerably less expensive and reliably provide safe drinking water. Hence, it becomes beneficial to treat water on the consumers' end. PoU water treatment systems can serve as a cost-effective solution to the risk of waterborne diseases. Most importantly, the PoU water treatment systems have a lower risk of water recontamination, thereby reducing the risk of waterborne diseases. Although most commercially available PoU water treatment systems require electricity or other resources, low-energy alternatives have also been developed.

7.3 COST-EFFECTIVENESS OF POINT-OF-USE WATER TREATMENT SYSTEMS

There are multiple studies available that measured a reduction in disease burden due to PoE and PoU water treatment systems. The growing market of membrane-based and other PoU water treatment systems creates opportunities for the water technology suppliers, particularly in developing countries, due to compromised water quality being supplied to the households. Thus, PoU water treatment systems remain a major solution for several reasons: poorly protected water sources, inefficient water treatment plants, leaking water distribution systems, intermittent water supply, and inadequate water safety in households, particularly in developing countries.

Various systematic reviews have indicated that household PoU systems can improve public health substantially. Meta-analyses of water, sanitation, and hygiene interventions estimate that PoU water treatment systems can improve the drinking water quality by 30%–40%, thereby reducing the users' susceptibility to waterborne illnesses. PoU water treatment systems have been extensively employed for decades to improve drinking water quality at the consumer's end. Several studies highlight the disease reduction associated with PoU water treatment. Verhougstraete *et al.* (2020) found that the disease burden caused by waterborne diseases is around 24.2 billion USD/year. The same study also found that the cost of installing PoU water treatment systems in the disease-prone households would be an 11.3 billon USD/year. Consequently, the healthcare cost would be reduced by 928 million USD (Verhougstraete *et al.*, 2020). Similarly, a study funded by Water Quality Research Foundation (WQRF) assessed the PoU water treatment system's effectiveness for lead removal. The cost–benefit analysis found that for every 1 USD spent to reduce lead hazards, there is a savings of 17–220 USD to society. According to the study, the healthcare costs needed to treat lead exposure are 435 000 000 USD to the community. In comparison, installing a PoU water treatment system would only cost 11 100 000 USD (WQRF, 2021).

Rogers *et al.* (2019) evaluated the cost-effectiveness of three PoU water treatment technologies (chlorine tablets, flocculent disinfection and ceramic filters). According to the study, chlorine tablets are the most cost-effective alternative, with a cost-effectiveness ratio of 24 USD, followed by flocculent disinfection (149 USD) and ceramic filters. In 2013, Ren *et al.* conducted a study evaluating the sustainability and cost-effectiveness of ceramic filters. According to the study, Ceramic PoU technology is less expensive per family unit (FU) than a centralised water treatment system, but it also has higher cost-effectiveness when diarrhoea is reduced. The cost of the centralised water treatment system per 1 FU was 221 USD, whereas, for the ceramic filter, the cost of delivering 1 FU was approximately 63 USD. These results indicate that the ceramic filters-based PoU water treatment systems are 3–6 times more cost-effective than the centralised water treatment system, and PoUs also exhibit better environmental performance (Ren *et al.*, 2013). Holmes *et al.* (2010) investigated the preventive efficacy and cost-effectiveness of filtration-based PoU water treatment systems in a subacute care unit and found that the total patient care costs were reduced by 248 136 USD after installing PoU water treatment systems, and the care unit saved 17 100 USD. Overall, it is apparent from the studies that PoU water treatment systems can be an effective measure for preventing waterborne diseases.

7.4 PoU WATER TREATMENT SYSTEMS MARKET DYNAMICS

7.4.1 Driver: increasing water contamination

Water pollution occurs when pollutants are dumped into water, directly or indirectly, without enough treatment to remove hazardous substances. Human activities' most frequent water pollutants include microbial diseases, nutrients, heavy metals, persistent organic matter and suspended sediments, pesticides, and oxygen-consuming compounds. Heat can be a contaminant since it elevates the temperature of the water. Pollutants are the most common cause of significant water quality impairment. To decontaminate this water, various processes such as purification, disinfection, and chemical treatment are done before being supplied to various residential and commercial units for consumption. However, some pathogenic microorganisms and inorganic compounds remain in the water that PoU water treatment systems remove. Also, these water pollutants are increasing daily, and the water that needs to be treated boosts the PoU water treatment systems market.

7.4.2 Restraint: high installation, equipment and operational cost

Despite the numerous benefits of water treatment, the cost of installing PoU water treatment equipment is prohibitive. A water softener, for example, can cost between 2000 and 4000 USD to transform hard water into soft water. Installing a water softener necessitates the use of professionals and installers, resulting in significant installation expenses. Some of the PoU water treatment systems have substantial operating and maintenance expenses. Distillation systems, for example, utilise a significant amount of energy for both cooling and heating. Some Asian and African countries are particularly affected by these issues. Water distribution and storage infrastructure are also lacking in underdeveloped Asian and African countries. This could be a stumbling hurdle when installing PoU water treatment.

7.4.3 Opportunity: scarcity of clean water in developing and underdeveloped countries

Water scarcity affects some Asia Pacific countries, such as Bangladesh, Pakistan, Nepal, and most African countries, Nigeria, Ghana, and Ethiopia. Because these countries are poor and undeveloped, they lack the financial stability to provide clean drinking water (www.marketsandmarkets.com/Market-Reports/point-of-use-water-treatment-systems-market-131277828.html). Diarrhoea, which is caused by contaminated drinking water, is responsible for 8.5% and 7.7% of all deaths in Asia and Africa, respectively, according to the WHO. According to the report, Argentina, Bangladesh, India, Chile, and Mexico drink water tainted with arsenic, which causes skin damage.

7.4.4 Challenge: ageing infrastructure

The majority of wealthy countries are confronted with the issue of ageing infrastructure. These countries constructed water treatment facilities decades ago and continue to employ outdated technology. Large sums of money will be required to rebuild and upgrade these ageing infrastructures. According to a US-based PoU water treatment system manufacturer, approximately 384 billion USD is needed to replace the existing water infrastructure. It further stated, 'America's water infrastructure is nearly at the end of its useful life'.

7.4.5 Counter-top units are the widely preferred device of point-of-use water treatment systems

Counter-top units are projected to be the largest segment in the PoU water treatment systems market. Counter-top units are also known as on-counter filters. These filters sit on the counter and are directly connected to the faucet. They consist of a diverter, allowing users to switch between unfiltered and filtered water. Counter-top units operate on RO as well as activated carbon technologies. These filters reduce contaminants, such as bacteria, dirt, chlorine, particulates, rust, lead, mercury, sediment, copper, benzene, cadmium, and cysts. The significant advantage of counter-top units is that they do not require frequent filter changes. However, they require some plumbing. Counter-top units do not chill water, unlike pitcher water filters.

7.4.6 Increased demand from the residential sector

The residential sector is projected to be the largest PoU water treatment systems market segment by application. The residential application mainly encompasses PoU water treatment systems to produce potable water for domestic consumption. The residential water treatment application is expected to witness high growth due to the increasing need for treated drinking water, removing unpleasant taste, odour, discolouration, suspended solids, biodegradable organics, and pathogenic bacteria.

7.4.7 Reverse osmosis is the most preferred point-of-use water treatment technology

RO is projected to be the largest PoU water treatment systems market segment by technology. RO is a process of producing pure water through a semipermeable membrane, wherein the water is pumped at high pressure through this membrane, which separates inorganic minerals (such as radium, sulphate, calcium, magnesium, potassium, sodium, nitrate, fluoride, and phosphorous), organic compounds (including pesticides), and other impurities from water. Usually, RO systems are combined with a mechanical and activated carbon

Table 7.1 Country-wise cost of PoU water treatment systems.

Country	Cost (in USD)
India	130–326
Pakistan	110–140
United States of America	200–1500
Canada	150–800
England	581–600
New Zealand	100–325
China	60–500
Japan	241–600
Germany	40–120
France	50–130
Spain	28–50
Italy	42–60
United Arab Emirates	50–440
South Korea	30–70

filter. The sand and large particles are removed by the mechanical filter first, and then after passing through the RO unit, the water is moved through the activated carbon filter to remove organic compounds. The RO water purifier can be of various types, such as wall-mounted or tabletop and under-the-sink or under-the-counter RO purifiers.

7.5 COUNTRY-WISE COSTS OF POINT-OF-USE WATER TREATMENT SYSTEMS

The global PoU water treatment systems market size is projected to grow from 16.2 billion USD in 2021 to 25.3 billion USD by 2026, at a compoundannual growth rate (CAGR) of 9.3% from 2021 to 2026. The PoU water treatment systems market is expected to witness significant growth due to the increasing demand for clean drinking water, increasing water contamination, growing human population, growing awareness about the benefits of water treatment, and technological innovations in the water treatment industry. The prices have been listed as per the popular e-commerce websites in the countries for various PoU water treatment systems, as presented in Table 7.1.

7.6 ECONOMIC ANALYSIS OF POINT-OF-USE WATER TREATMENT SYSTEMS

The PoU water treatment system market is affected by several trends, including population growth and water pollution. It is likely to be also influenced by water stress based on climate changes. The term economics refers to evaluating capital costs and operating costs associated with the fabrication/manufacturing of the

PoU water treatment system. The kind of analysis employed and the values provided to the selected economic factors determine the economic assessment of a certain separation procedure. As a result, economic assessments given by different evaluators may differ significantly. However, if the approach employed in economic analyses is explicitly disclosed, such disparities can be instructive.

PoU water treatment systems treat water used for drinking and cooking. The quantity of water required for drinking and cooking has considerable variation primarily based on environmental conditions and is reported to be varying from 2 L per person per day for drinking water to 8 L day^{-1}. PoE water treatment system normally treats the entire water supply to the individual household. Water treatment technologies available for conventional water treatment plants can also be used for PoU and PoE water treatment systems. In addition, other treatment technologies are also used for PoU and PoE water treatment systems. Some of the most popular PoU water treatment technologies, which are also referred to in earlier chapters, are given below:

- Sedimentation or settling
- Filtration (media-based, ceramic, membrane)
- Coagulation, flocculation, and precipitation
- Adsorption
- Electro-chemical
- Ion exchange
- Chemical disinfection
- Solar disinfection
- UV disinfection

There has been substantial growth in the PoU water treatment systems market in the last 10 years, and it is expected to grow further from over 34 billion USD in 2028 to 19.8 billion USD in 2021. Awareness about waterborne diseases, the erratic water supply of compromised water quality, unsafe water handling and gradual reduction in the cost is driving the global market of PoU water treatment systems. Although demand for PoU water treatment systems is increasing globally, Asia holds the largest market share. It is likely to remain the fastest-growing continent at a CAGR of about 10% from 2020 to 2030. Increasing awareness about safe water, rapid urbanisation and affordability of PoU systems in Asia are creating substantial growth opportunities for PoU water treatment systems. Almost 35% of the world's urban population resides in Asia. China, Japan and India are major consumers of PoU water treatment systems, and local production facilities cater to the domestic demand in these countries. India is expected to register a maximum growth rate of PoU water treatment systems sales. Pacific countries such as Australia also have very high adoption of PoU water treatment systems other than Asian countries such as Japan, South Korea and so on. There is increasing demand for PoU and PoE systems in the United States of America due to the availability and penetration of such water treatment systems. The demand is also accelerated due to major commercial manufacturers of these systems with well-established distribution channels in the region.

The PoU and PoE water treatment systems market is dominated by key global players who can provide the systems through their national counterparts. These multinational companies invested considerable resources to develop PoU water treatment systems and products mainly driven by consumers' psyche rather than demand for the new products. A diverse strategy involving a broad range of product portfolios, technological improvement, catering to variable water quality parameters, ease of operation and maintenance, and cost-effectiveness is used for new product development. However, many local companies in China, India and other emerging economies are recording their presence due to the availability of cheaper resources and relaxed regulatory regime. Due to comparatively lower costs, several PoU water treatment systems manufactured in these countries are also imported.

However, these systems remain unaffordable to many consumers, particularly in developing countries which is the major limiting factor to further improving demand. Membrane-based PoU water treatment systems, particularly RO, are reported to be more than half of the systems sold globally, which clearly indicates a preference for such systems.

Data related to the cost of PoU water treatment systems are available. However, there is a large variation in the cost of the system depending on the year of reported data, type of technologies and country of origin. Moreover, the capacity of PoU water treatment systems is also not known to determine and effectively compare these systems. The approximate cost of water treatment units for various PoU water treatment system categories is presented in Table 7.2 by using two important references. Costs of these systems should be used for comparing different technologies having similar capacities. Membrane processes are the most expensive among other alternate technologies. However, these technologies are not comparable with membranes-based technologies in terms of contaminant removal potential. The cost of PoU water treatment systems presented in Table 7.2 is very high as the cost of these systems has considerably reduced over the years. These data can be used for comparing the cost of the systems with each other.

The cost mentioned in Table 7.2 does not include one-time labour costs (~300–500 USD) or the costs of any electrical power required. Following is the summary of the cost of various PoU water treatment systems:

- PoU water treatment systems can be employed to treat water at the consumer end. The costs of PoU systems vary depending on the technology it utilises.
- Technologies like solar disinfection can serve as the cheapest alternative for water disinfection since it requires only sunlight and transparent bottles. The fabrication costs of solar disinfection units are ~40–50 USD, which is much lower than the fabrication costs of some other PoU water treatment systems.
- Size-exclusion technologies are also very cost-efficient, with ceramic filters having the lowest cost, followed by the activated carbon filter. The cost of UF membrane is currently around 40 USD m^{-2} and declining, which can be very beneficial for developing countries.

UV-based PoU water treatment system is also a commonly employed technology. UV-based PoU water treatment system requires electricity which

Table 7.2 Approximate costs of PoU water treatment systems.

PoU/PoE Technology	Average Cost (Materials Only)	Reference
UV disinfection	800–2000 USD	www.fixr.com/costs/ water-purification-system
Ionisation	1000–2000 USD	
Distiller	1200–4000 USD	
UF/RO	500–1500 USD	
Activated carbon block filter	50–500 USD	
Activated granular carbon filter	50–500 USD	
Sediment filter	85–1000 USD	
Activated alumina	200–2000 USD	
Boiling with fuel	Depends on fuel price (+pots required for boiling)	Peter-Varbanets *et al.* (2009)
Solar disinfection	None (plastic bottles required)	
UV disinfection with lamps	100–300 USD (+operational costs 10 USD–100)	
Free chlorine	2–8 USD(+operational costs: 1–3 USD)	
Biosand filters	10–20 USD	
Ceramic filters	10–25 USD(+operational costs: none to 10 USD)	
Coagulation, filtration, chlorination	5–10 USD (+operational costs: 140–220 USD)	
Activated carbon filtration	25–50 USD (+operational costs 25–50 USD)	
Microfiltration	3 USD (+operational costs: 12 USD)	
UF	40 USD	
RO	300 USD–600 USD (+operational costs: 80 USD–20 USD)	

adds to the operational costs. Hence, even though UV-based PoU water treatment systems may cost less than filtration systems, their operating costs are relatively higher. A study conducted by Verhougstraete *et al.* (2020) measured the cost-effectiveness of multiple PoU water treatment systems. It was found that over 5 years, costs of operations were 680 USD for RO-based PoU water treatment system, 546 USD for activated carbon filtration, and 645 USD for pour-through filtration units per household. Similarly, UV-based PoU water treatment system costs 1500 USD, absorptive media filters cost 936 USD, and distillation costs 740 USD. The ion exchange-based PoU system was found to be the most expensive option, costing 1870 USD. The cost of various PoU systems/ technologies for removing nitrate and perchlorate and their associated units

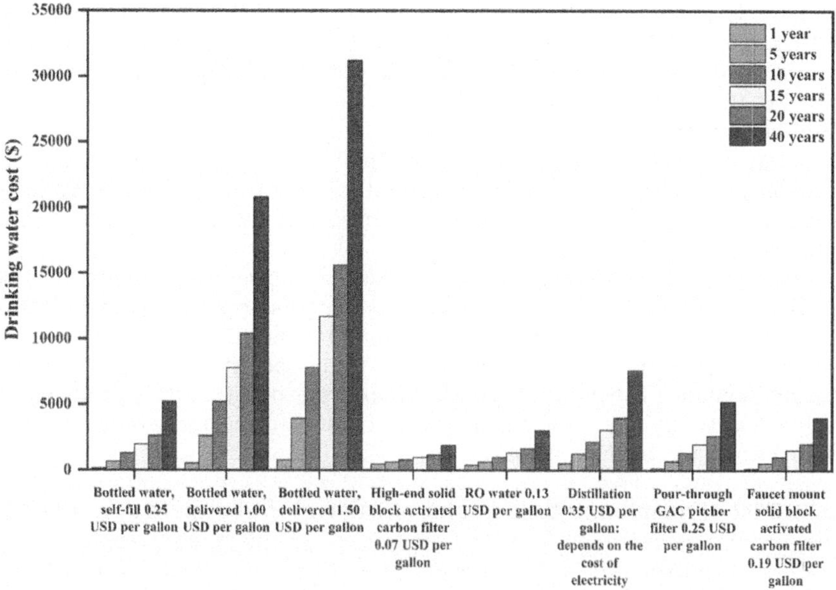

Figure 7.1 PoU technologies and associated costs.

is presented in Figure 7.1 (www.cyber-nook.com/chart/default.asp, 2018). The data are based on 38 L of water used per person per week, 5.3 L of water used per person per day and 1968 L water used per person per year.

For extremely small systems (below 200 households), PoU technologies (RO) have the lowest costs for both nitrate and perchlorate. For larger systems, anoxic biological treatment systems have the lowest costs, although, for perchlorate, low concentrations and the high capacity of the selective resins favour ion exchange. Higher influent concentrations favour biological treatment.

Capital cost in manufacturing PoU water treatment systems should theoretically include the cost incurred for the facilities, including land costs. Providing manufacturing/capital costs of the facilities is beyond the scope of this book. Moreover, capital investment in PoU water treatment system procurement is very small compared to large desalination plants. Considering the PoU water treatment system is independent of manufacturing facilities, the capital cost can be considered as the procurement cost of the system.

7.6.1 Capital investment

This sums up all of the costs incurred at the start of the PoU water treatment system's manufacturing facility (plant) life. Capital investment can be estimated using several ways. The amount of specific information available and the level of precision desired to determine which method is being used. Several approaches are described elsewhere. Direct production costs normally consider expenses directly associated with the manufacturing operation, such as expenditures for raw materials, supervisory and operating labour, plant maintenance and repairs, operating supplies, royalties, and so on. Advertisement expenses also form a major component in branding PoU water treatment systems and also added to the expenses of manufacturing of these systems. This book does not describe the method to determine the direct production cost of PoU water treatment systems based on the expenses of manufacturing facilities. Moreover, considering that the market currently is full of PoU water treatment systems from numerous manufacturers, competitiveness also decides the cost of these systems.

In the case of the PoU water treatment system, the capital investment from the consumers' perspective can be considered as the procurement and installation cost of the unit. A cost estimate includes more than just the capital investment. Another crucial component is estimating the unit's operational costs. Operating and maintenance expenditure (OPMEX) covers expenses for the consumers, such as electricity operating supplies and component replacement in the case of PoU water treatment systems.

7.6.2 Cost of membrane-based point-of-use water treatment systems

The membrane-based PoU water treatment systems are presented as a new, advanced solution to the consumers and have the potential to make a motivational appeal, which is getting increasingly important for the adoption of the PoU water treatment system. Another factor that may increase adaptability of membrane-based systems for low-income consumers is the desire for high-tech options to have class sensitivity.

The demand for membrane-based PoU water treatment systems is increasing for several reasons. These systems have become synonymous with household water treatment systems. The majority of the literature related to the costing of membrane-based water treatment systems is available for large desalination plants.

7.7 GLOBAL MARKET SIZE OF MEMBRANE-BASED POINT-OF-USE WATER TREATMENT SYSTEMS

Substantial growth in water treatment systems fuelled by growing urbanisation has led to increased demand for membrane technology across the globe. The supply of potable drinking water to households is one of the topmost necessities for all countries. The global domestic water consumption rate increases by about 100% every 20 years. The rising scarcity of drinking water, increasing population and growing water demand for other sectors are the major drivers of the demand for membrane-based water and wastewater treatment market throughout the globe. Several reports analyse the growth of membrane industries, particularly its application in water and wastewater treatment. These reports and data analysis of membrane-based water and wastewater treatment systems indicate the range of market size and CAGR. However, all these reports provide a clear indication of substantial growth in this type of market. As per one such available report, the global membrane-based water and wastewater treatment systems market is estimated to be 13.5 billion USD in 2019 and projected to grow to 19.6

Figure 7.2 Membrane water and wastewater treatment market growth rate by region (2020–2025).

billion USD by 2025, with a CAGR of 6.4% during this period. The membrane market is dominated by RO membranes, which are increasingly used in PoU water treatment systems. The growth of membrane-based water and wastewater treatment technologies is shown in Figure 7.2 (www.mordorintelligence.com/industry-reports/membrane-water-and-wastewater-treatment-wwt-market).

RO-based PoU water treatment systems are also very common in the global market, with an estimated cost of 300–1500 USD. Compared to RO-based PoU water treatment systems, the expenses of UF and microfiltration are much lower, with lower operational costs as well.

7.8 COST DETAILS OF MEMBRANE-BASED POINT-OF-USE WATER TREATMENT SYSTEMS: A CASE STUDY FROM INDIA

India is the seventh-largest country having a population of 1.3 billion. India has about 16% of the world's population, with only 4% of water resources. Hence, water resources are under stress due to a large population, industrialisation and agricultural activities. Water quality is also affected due to sewage, agricultural runoff and industrial discharges; quality improvement is reported progressively. With the ambitious programme, Jal Jeevan Mission, to provide safe drinking water to every household in India, improvement in services and water quality is expected.

India is rapidly growing marked for PoU water treatment systems, and membrane-based PoU constitutes a major component. India's PoU water treatment systems are dominated by use in households, which are mainly based on the RO. Membrane-based PoU water treatment systems are mostly distributed and sold through outlet stores, whereas UV and membrane (RO)-based water treatment systems are sold through the 'direct to home' channel

Table 7.3 Cost of commercially available membrane-based PoU systems in India.

PoU Unit	Technology Used	Rate of Production (L h^{-1})	Storage Capacity (L)	Cost (INR)	Additional Energy Requirements (+/−)
Aquaguard Enhance	RO+UV	12–15	6–10	21 000	+(∼35 W)
Aquaguard Superb	UV+UF	RO path: 12–15 UV path: 35–40	6	25 000	+
Aquaguard Classic	MF+UV	120	No storage	12 000	+(∼8 W)
Tata Swach	MF	4–5	7.5	1500	NA
Tata Swach Viva	UV+UF	30	6	10 000	+
Kent Pearl	RO+UV+UF	20	8	20 500	+(∼60 W)
Kent Prime Plus	RO+UV+UF+UV in storage tank	20	9	21 000	+(∼60 W)
HUL Pureit Classic	RO+MF	10–12		9000	+(∼36 W)
Pureit Marvella	UF+UV	24	10	17 500	+
Bluestar Iconia	RO+UV	60		10 000	+

of about 30%. As a result, many well-known key players have developed their own exclusive branded outlets for cell membrane (RO)-based water treatment systems. Most new and current companies concentrate on membrane-based water treatment systems, particularly RO-based ones, which are the fastest-growing application in the Indian PoU water treatment systems market. The membrane (RO)-based water treatment systems market is largely consolidated, with 10 global companies accounting for over 70% of the market share. New domestic and international players have also entered the market. UF is a water treatment technology that is increasingly gaining traction in India as an alternative to RO. The costs and technical specifications included in Table 7.3 are per the manufacturer's websites.

RO and UV filters are the most commonly used technologies in the Indian market. The average cost of commercially available PoU devices in India is INR 15 000 (USD 200). Most PoU water treatment systems require electricity and incur additional operating and maintenance costs.

7.9 SUMMARY

The PoU water treatment systems are much more feasible since they are considerably less expensive and provide safe drinking water. Although conventional water treatment plants and now water safety plan approaches are meant for ensuring safe water to the consumers, their efficient application is limited largely to developed countries. Hence, it is beneficial to treat water on the consumers' end. PoU water treatment systems can serve as a cost-effective solution to the risk of waterborne diseases. The cost-effectiveness of PoU water treatment systems is well documented. Membrane-based PoU water treatment systems are increasingly being used and recording impressive growth globally due to the versatility of such systems in treating many contaminants in drinking water. RO membranes form a major proportion and are often synonyms for PoU water treatment systems. Considering the reduction in the cost of membrane-based PoU water treatment systems, many variants exist in countries such as China and India. With the increase in water scarcity and questionable water quality on consumers' end, the use of PoU water systems in general and membrane-based PoU water treatment systems, in particular, will continue to grow in the next decade.

doi: 10.2166/9781789062724_0167

Chapter 8

Certification and evaluation of membrane-based point-of-use water treatment systems

8.1 INTRODUCTION

Water-borne disease outbreaks still occur globally despite the continued efforts of numerous government and non-government entities to maintain water safety. PoU water treatment technology can be a temporary but immediate solution. Meta-analyses of various studies have found that some household PoU water treatment systems can improve drinking water quality and reduce the risk of infections by 30%–40%. Given the wide range of technologies available for PoU drinking water treatment processes, it is critical to creating quality standards that help consumers choose the right system.

Due to natural or anthropogenic pressures, deterioration in water quality has resulted in an opportunity to innovate and roll out technologies/products that can deliver safe water to households. These innovative technologies/ products, including membrane-based, aim to treat the contaminated water to adhere to prescribed water quality standards/guideline values and provide an interim solution for improved quality of safe and reliable drinking water. Apart from directly using these PoU water treatment systems by the consumers, many governmental/non-governmental agencies also provide these systems through national policies and health programmes for technological interventions. Development, procurement and distribution of water treatment products worldwide are continuous.

Advancements and innovations have occurred in early-developed technologies such as adsorption, ion exchange, membrane separation, disinfection, etc. Every passing year, several PoU water treatment systems are developed and introduced in the market with various configurations to address the supply of clean water at the PoU with varying and sometimes unverifiable claims. With the market full of such products and huge claims, it is essential to understand the complexities of water treatment and evaluate these claims independently based on the common testing/evaluation/certification protocol.

The PoU water treatment products/units/systems evaluation and certification guides consumers, including private and government agencies, in selecting suitable systems and help the national government directly or indirectly in several technical- and evaluation-related functions. The primary objective of certification/evaluation is to provide an independent third-party evaluation of PoU water treatment systems and certify them. Additional objectives of certification/evaluation can be as follows:

- To create a platform to promote and coordinate the independent and reliable evaluation of PoU water treatment products/units/systems based on the uniform protocol and standards.
- To provide a common platform for technology providers and manufacturers to assess the performance and efficacy of their technology.
- To aid the government/non-government agencies which are involved in the procurement of PoU water treatment products/units/systems for the public at large to ensure access to safe drinking water.

Many organisations and governments now provide suggestions, guidance, protocols, specifications, and processes, or descriptions of methods and materials for evaluating the effectiveness of PoU water treatment systems. The WHO, Water Quality Association (WQA), National Science Foundation (NSF)/American National Standards Institute (ANSI), and the United States Environmental Protection Agency (USEPA) are among the most prominent of these; all of which now offer certification or some other type of recommendation based on performance appraisal. Recognising the importance of assisting countries in formulating and implementing nationally oriented household water treatment and safe storage policies and programmes for raising awareness and driving actions, such initiatives must continue to be encouraged and supported. These agencies worldwide have or are establishing administrative and technical processes that allow for country/state-specific registration and use of PoU water treatment systems/technologies/products. For that instance, PoU water treatment systems/technologies/products certification/approval/evaluation can be limited to a single country, recognised regionally or internationally.

8.2 STANDARDS FOR EVALUATION/CERTIFICATION FROM INTERNATIONAL AGENCIES

PoU water treatment systems/technologies/products performance evaluation and certification or approval programmes are already in place at the WHO, NSF/ANSI and WQA. Currently, several organisations provide specifications, standards, recommendations, evaluation protocols, and procedures for PoU water treatment systems. WHO, NSF-International and the USEPA are the foremost among these organisations providing certifications and approvals after the formal evaluation of PoU water treatment systems. Other organisations, such as Underwriters Laboratory (UL) and the International Association of Plumbing and Mechanical Officials (IAPMO), issue certifications based on these publicly available materials. While not necessarily comprehensive or representative of policies, procedures, and methods utilised worldwide, they reflect the most well-known international information resources and certain countries with their systems (Bailey *et al.*, 2021).

Traditionally, regulatory bodies have taken a harder stance against PoU treatment devices than those against PoE systems. However, a closer examination of the chronology in Figure 8.1, which displays the evolution of attitudes towards PoU water treatment systems, reveals a steady shift in the regulatory consideration of PoU systems. As mentioned in the next subsections, several water regulations have accepted PoU water treatment systems as an alternative to meet maximum pollutant levels (Hamouda *et al.*, 2010). Table 8.1 summarises the overview of standard protocols for evaluating PoU water treatment systems.

8.2.1 World Health Organisation

According to a consensus publication by WHO, 'Evaluating household water treatment options: Health-based targets and microbiological performance specifications' (WHO, 2011), three specific levels of technical microbial reductions are recommended for PoU water treatment systems. This publication also mentions the development of local application performance specifications to protect consumers and inform decision makers to select appropriate technologies or approaches. This publication provides a basis to evaluate the microbiological performance of the PoU water treatment system to

- establish health-based performance targets, for example interim target, protective and highly protective to have an incremental improvement in water quality
- guide the development and strengthening of testing protocols

Performance targets for three classes of microorganisms, that is bacteria, viruses and protozoa, are included in this publication. The reference pathogens used in this publication are *Rotavirus* for viruses, *Campylobacter jejuni* for bacteria, and *Cryptosporidium* for protozoan parasites. They are comparatively better characterised, highly significant for public health importance, and conservative in relation to infectivity and dose–response. Alternatively, if PoU water treatment systems were operational to remove these pathogens under reference, other pathogens within each category of pathogens would also be removed. The quantitative microbial risk assessment (QMRA) approach is used to develop performance targets for these pathogens. The publication includes material and methods, recommendations, and processes for performing technological evaluations that may be contextualised to other countries or standard development agencies' bodies. To a large extent and keeping practicality in view, the publication also promotes the adaption of available testing methods. The recommendations made in this publication do not apply to chemical contaminants.

The WHO has established an 'International Scheme to Evaluate Household Water Treatment Technologies' to evaluate the microbial performance of PoU water treatment technologies against WHO's health-based criteria. The scheme

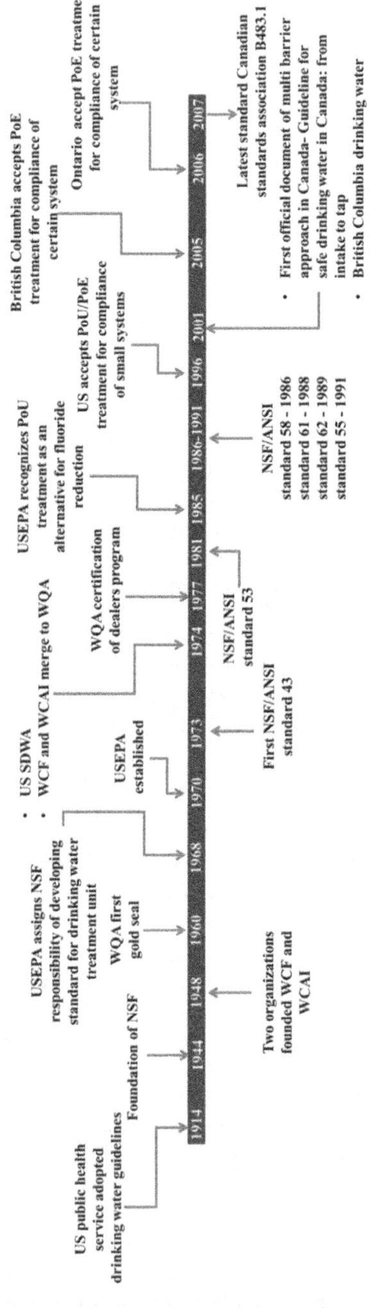

Figure 8.1 Timeline for the evolution of governance and management practice of PoU/PoE water treatment system.

Table 8.1 Overview of standard protocols for evaluation of PoU water treatment systems.

Standard Protocol	Technologies Covered	Reduction Requirements			Methods Described
WHO harmonised testing protocol	Microbiological purifiers and disinfection technologies	*Highly protective* Bacteria: $\geq$4 log Viruses: $\geq$5 log Cysts: $\geq$4 log	*Protective or limited protection* Bacteria: $\geq$2 log Viruses: $\geq$3 log Cysts: $\geq$2 log	*Interim* Achieves two 'protective' classes	Testing methods not described. Calculation methods (for log reduction and geometric mean) are described.
NSF/ANSI 42	Carbon filters and media filters	TOC reduction: 2 mg L^{-1} reduced by $\leq$50% during a 50% ON/OFF cycle Chloramine reduction: 3.0–0.5 mg L^{-1} during a 50% ON/OFF cycle Bacterial reduction: 1–6 log reduction. The geometric mean of effluent should not be higher than influent over 6–13 weeks, 2% ON/98% OFF cycle Iron (Fe^{2+}): 3/5–0.3 mg L^{-1} Manganese reduction: 1/2–0.05 mg L^{-1} Zinc reduction: 10–5 mg L^{-1}			Testing methods for GC–MS analysis protocol, chemical reduction testing protocols (e.g., monochloramamine reduction tested using DPD ferrous titrimetric method or the DPD colorimetric method), and heterotrophic plate count method for microbial reduction is described.
	Sediment filters	Particulate reduction 85% reduction class: I ($\geq$0.5 to <1 μm), II ($\geq$1 to <5 μm), III ($\geq$5 to <15 μm), IV ($\geq$15 to <30 μm), V ($\geq$30–50 μm), VI ($\geq$50 to <80 μm)			Mechanical reduction testing protocols. Standard particle counting technologies as available from the particle counter used.

NSF/ANSI 53	Carbon and media cartridges	VOC surrogate (CHCl$_3$): 51 VOCs, acceptable limit 0.30–0.015 mg L^{-1} Metal reduction: AsIII, IV: 0.05/0.3 to <0.01 mg L^{-1} Ba: 10–2 mg L^{-1} Cd: 0.03–0.005 mg L^{-1} Cr: 0.3–0.01 mg L^{-1} Cu: 3–1.3 mg L^{-1} Se: 0.10–0.05 mg L^{-1} Hg: 0.006–0.002 mg L^{-1}, Pb: 0.15–0.01 mg L^{-1} Fluoride: 8–1.5 mg L^{-1} NO$_3$+NO$_2$–N: 30–10 mg L^{-1}	Chemical reduction testing: Standard methods used; USEPA method 200.8 for Hg, Se, Pb, Ba, Cd, and Cr testing or 200.9 for Pb testing.
	Mechanical filtration	Turbidity reduction: 10 NTU to 0.5 NTU up to 75% flow reduction Cyst reduction (3-μm microspheres): 99.95% (~4 log)	Turbidity analysis: USEPA method 180.1 Cyst analysis: Cyst detection and enumeration methods using UF and DAPI staining
NSF/ANSI 55	UV-based drinking water treatment systems	MS2 phage reduction for class A system: Challenge level 10^4–10^5 pfu mL^{-1}; 4–5 log reduction should be observed. *Saccharomyces cerevisiae* for class B system: 4–5 log reduction at 104–105 CFU mL^{-1} influent challenge. 2.14 log reduction with UV source irradiance at 100% normal output.	Standard USEPA methods 525.2, 528, and 625 are suggested for chemical testing. Standard pour plate used for microbial enumeration.

(Continued)

Table 8.1 Overview of standard protocols for evaluation of PoU water treatment systems. (*Continued*)

Standard Protocol	Technologies Covered	Reduction Requirements	Methods Described
NSF/ANSI 58	RO technology	TDS reduction: 750 ± 40 of NaCl to 187 mg L^{-1} (75% reduction) Turbidity reduction: Influent challenge level of 11 ± 1 NTU reduced to at least 0.5 NTU. Metal reduction AsV, AsIII, Ba, Cd, Cr, Cu, Pb, and Se same as NSF/ANSI 53 F, NO$_3$, NO$_2$, VOC reduction NSF/ANSI 53 Cyst reduction 99.95% reduction	USEPA Method 625 (GC-MS method) is used for chemical contaminants (such as arsenic) detection. Other protocols are the same as described in NSF/ANSI 53.
NSF Protocol P231	UV, halogen, iodine-based treatment technologies, Size-exclusion treatment technologies like ceramic or membrane.	Microbial reduction: Bacterial reduction: 6 log reduction at influent challenge 10^7 CFU 100 mL^{-1} (*K. terrigena* used) Viral reduction: 4 log reduction at the influent challenge of 10^7 PFU L^{-1} (MS2 phage used) Cysts reduction: 3 log reduction at influent challenge 10$^{6-7}$ cysts L^{-1}. Live cysts of *C. parvum* or 4–6 μm microspheres	Testing should be done for 14½ days. Volume% tests eight sampling points. Standard pour plate method is used for detection and enumeration.
WQIA IP-100 guide standard and protocol for microbiological evaluation of drinking water treatment devices	UV, halogenated, iodinated, membranes size-exclusion technologies	Microbial reduction: Bacterial reduction: 6 log at influent challenge of 5×10^7 CFU 100 mL^{-1} (*K. terrigena* used) Viral reduction: 3 log on day 5 and 4 log on day 6 at the influent challenge of 5×10^7 pfu/100 ml. Cyst reduction: Not required. Alternatively, 2 log bacteria and 3 log viral (only at 60% estimated capacity sample) with 3 μm Microspheres (cyst) reduction.	Standard pour plate and standard membrane filtration method for bacterial reduction. Immunofluorescence techniques used for viruses.

is to help make informed decisions about the selection of PoU water treatment systems by the countries, international organisations and other agencies. Many treatment technologies are available to remove pathogens from drinking water, and many manufacturers and vendors provide the products/systems based on these technologies. However, the performance of these treatment technologies is variable in terms of removal of the pathogens, namely bacteria, viruses and protozoa posing health risks.

The WHO constituted an Independent Advisory Committee (IAC) to provide technical and scientific advice to develop and operationalise the evaluation scheme. Experts from the subject field of microbiology, water quality, and water treatment technology formed the IAC. The IAC periodically advises on selecting testing laboratories, harmonised protocols for evaluation and testing, reporting templates of test results, and analysis of test findings for the PoU water treatment systems submitted for evaluation under the scheme. WHO developed recommendations to assist the countries and other users of the scheme in appropriately evaluating and selecting PoU water treatment technologies/systems. The guiding principles and criteria for evaluating these products/systems' performance are also detailed in this publication. These systems are classified into three levels of performance: 3-star (★★★); 2-star (★★), and 1-star (★), considering the performance in removing pathogens from treated drinking water, details of which are given in Table 8.2.

The publication does not include a detailed evaluation protocol. However, recommendations and guidance related to testing processes, materials and methods, and synthetic test water details are provided. The document suggests choosing the target organism based on the disease prevalence in the area. If there is insufficient data, the document recommends using *Escherichia coli* (for the bacterial challenge), MS-2 Phage or phiX-174 (for the viral challenge), and *Cryptosporidium parvum* (for infectious oocysts). Sample collection procedures have also been described in the document. The development of the evaluation protocols has been left up to the user. Rotavirus is recommended as an acceptable viral test candidate in the WHO document since it is the reference viral pathogen in the GDWQ, 2022. Alternatives to reference pathogens are also identified, including Echovirus 12 (an enterovirus), coliphage MS2, ϕX-174

Table 8.2 Performance criteria for PoU water treatment systems.

Performance Classification	Bacteria (log_{10} Reduction Required)	Virus (log_{10} Reduction Required)	Protozoa (log_{10} Reduction Required)	Interpretation (with Correct and Consistent Use)
★★★	≥ 4	≥ 5	≥ 4	Comprehensive protection
★★	≥ 2	≥ 3	≥ 2	
★	Meets at least 2-star (★★) criteria for two classes of pathogen			Targeted protection
-	First to meet WHO performance criteria			Little or no protection

(coliphage), and/or other bacteriophages. They are physically, morphologically, chemically, and behaviourally similar to other enteric viruses, providing a framework for evaluating the virus reduction performance of PoU water treatment systems (technologies).

8.2.2 National Science Foundation/American National Standards Institute

Water systems should ensure that any PoU or PoE water treatment system is properly certified/evaluated. ANSI has issued product standards (formerly NSF/ANSI standards) for a specific type of PoU or PoE water treatment systems. NSF/ANSI standards cover six types of PoU and PoE water treatment systems (NSF International):

- NSF Standard 42: Drinking Water Treatment Units – Aesthetic effects;
- NSF Standard 44: Cation Exchange Water Softeners;
- NSF Standard 53: Drinking Water Treatment Units – health effects;
- NSF Standard 55: UV Microbiological Water Treatment Systems;
- NSF Standard 58: RO Drinking Water Treatment Systems;
- NSF Standard 62: Drinking Water Distillation Systems;
- NSF Standard 244: Intermittent incursions or accidental microbiological contamination;
- NSF Standard 401: Drinking Water Treatment Units – Emerging Compounds/Incidental Contaminants.

It is essential to realise that some of these initiatives are designed to test technology for providing supplementary treatment to public or private drinking water supplies. The NSF/ANSI Standard 55 NSF International (2019b) and NSF P231 NSF International (2014) are related to treatment technology evaluation. At the same time, NSF/ANSI Standards 55 NSF International (2019b), 62 NSF International (2019a), 244 NSF International (2019c), and NSF/ANSI Protocols P231 NSF International (2014) and P248 NSF International (2012) evaluate performance for three classes of microorganisms (bacteria, virus, and protozoa). NSF/ANSI also has standards for chemical reduction performances (NSF Standards 53 NSF International (2019d) and 58 NSF International (2019e)), and only protozoa reduction is the only microbiological claim.

Independent entities that evaluate PoU water treatment systems/technologies/ products follow prescriptive protocols to fulfil specified performance levels for removing/inactivating viruses, bacteria, and protozoa (their surrogates are also used sometimes) that these agencies/organisations specify. The NSF/ ANSI P231 NSF International (2014) protocols assess performance as per the USEPA Guide Standard and Protocol for Testing Microbiological Water Purifiers (1987). There are provisions for non-performance-based items such as literature, material safety, and labelling on the systems included in NSF P231 NSF International (2014). For viruses (4 $\log_{10}$), bacteria (6 $\log_{10}$), and protozoa (6 $\log_{10}$), these standards/protocols contain very stringent and comprehensive testing performance parameters as well as numerical microbiological reduction performance requirements (3 $\log_{10}$). However, these organisations are aware that testing methodologies must be tailored to the attributes and characteristics of the technology being evaluated.

8.2.3 National Science Foundation-International

The NSF-International has a centre dedicated to evaluation/verification of drinking water treatment systems dedicated to technology verification since 2000. Consequently, the NSF-International has developed protocols for verification of the following specific water treatment technologies:

- NSF Protocol P231, 2014
- NSF Protocol P248 military operations microbiological water purifiers, 2020

These protocols complement the standards presented in the earlier section and should be read in conjunction with the standards. The protocols can be used as templates for developing test plans for evaluating/verifying individual PoU water treatment systems/technologies/products at specific locations. For example, NSF Protocol P231 addresses principles and procedures for verifying microbiological claims of water purifiers as recommended by the USEPA protocol. The NSF/ANSI 55 standard, as mentioned in the last section for UV Microbiological Water Treatment Systems, also has details similar to that in Protocol P231. Therefore, the test conditions and parameters differ from protocol to protocol, depending on the detailed specifications of the technology. The verification reports of technologies evaluated by NSF are publicly available.

8.2.4 Water Quality India Association

The Water Quality India Association (WQIA) is a non-profit organisation representing India's residential, commercial, and industrial water treatment industries. WQIA has developed a document providing guidelines for evaluating drinking water treatment devices based on the Bureau of Indian Standards (BIS) – IS 10500-2012, that is Indian standards for drinking water. This document is titled WQIA IP-100 guide standard and protocol for microbiological evaluation of drinking water treatment devices.

Procedures for performance evaluation of UV, halogenated and iodinated with size-exclusion technologies have been included in the document, along with the specific test conditions for each technology. Like the USEPA and WHO protocol, the WQIA protocol has also specified the reduction requirements of 6 log for bacteria, 3 log for viruses, and 4 log for protozoan oocysts. The protocol has also specified the reduction requirements that the PoU water treatment system should demonstrate according to its operational time (or its life point). The document is particularly useful since it provides detailed sampling, analysis, results recording, and PoU water treatment system conditioning procedures. The document also provides a template for a sampling plan for each technology. In short, the WQIA protocol has been developed specifically for Indian conditions and matches the global norms.

8.3 COUNTRY-SPECIFIC STANDARDS FOR EVALUATION/ CERTIFICATION

Many countries have also evolved protocols/guidelines/standards for evaluating PoU water treatment systems which, in some instances, are technology-specific.

Although they, in principle, are similar to various other protocols available in other countries or through international agencies, country-specific issues are also addressed. These protocols are briefly described below and summarised in Table 8.3.

8.3.1 Brazil

The Brazilian Standard NORM/ABNT/NBR 15176:2004 provides three separate categories for classifying the PoU water treatment system's performance efficiency. The PoU water treatment system needs to satisfy at least one out of the following three criteria for complying with this standard

- removal of free chlorine,
- particle retention efficiency,
- bacterial reduction ($\log_{10}$) according to the Brazilian standard.

The PoU water treatment system should exhibit at least 2 log reduction in water containing 10^7 colony forming units (CFU) 100 mL^{-1} of *E. coli*.

This methodology identifies and specifies a PoU water treatment system's performance efficacy based on particle retention efficiency, efficiency in lowering free chlorine content, and a minimum 2 $\log_{10}$ reduction that appears to be based on a health risk strategy. This performance target identified in the protocol has no evident rationale. Because the main microbiological performance criteria is a 2-$\log_{10}$ decrease of *E. coli*, whether by physical exclusion such as filtration or microbial inactivation methods, this strategy does not appropriately address microbial variety and variability. The changes in performance efficiency predicted to occur with viruses removed/inactivated/treated by filtration and chemical disinfection and protozoa treated by chemical disinfection are not effectively accounted for in this methodology and its microbial performance specifications.

8.3.2 Canada

Environment Canada prepared the Canadian protocol for performance testing of drinking water treatment technologies in collaboration with the Bureau de normalisation du Québec for the Canadian environmental technology verification program in 2015. The primary objective of the protocol is to assess the operational reliability of the technology. The duration of testing has been specified as a minimum of 12 consecutive months. Operating parameters have also been defined, and the number of samples need to be collected. Critical raw water parameters that need to be measured, such as pH, TDS, turbidity, total organic carbon, alkalinity, sodium, fluorides, iron, and manganese content, tri-halo methane, and halo acetic acid (HAA) values, have also been described. The document has also included total coliform and faecal coliforms in the testing parameters. Methods of detection of microorganisms have been described briefly, but the exact protocols have not been included.

8.3.3 China

The protocol 'GB/T 5750-2006 Standard' available in China through the Ministry of Health is used to evaluate PoU water treatment systems. The China

Table 8.3 Country-wise protocols for evaluation of PoU water treatment system.

Standard Protocol	Technologies Covered	Reduction Requirements	Methods Described
Indian standard IS14724:1999 (RA2009)	UV radiation technology	Microbial reduction: 3 log reduction at the influent challenge of 10^6 CFU 100^{-1} mL (*S. lutea* used) Chlorine reduction: After 50 L of filtration, 2–0.2 mg L^{-1} Turbidity reduction: After 50 L of filtration, 20–25 to 5 NTU	For microbial testing, the Standard membrane filtration technique (IS10500) used. Protocols for testing chlorine and turbidity reduction have not been described
Indian standard IS16240:2015	RO	TDS reduction: 1500–375 mg L^{-1} (75% reduction) Microbial reduction: 6, 4, 3 logs for bacteria (*E. coli*), viruses (MS2 phage), and cysts (*Cryptosporidium*), respectively Metal reduction: 3–60 times, as applicable, of IS10500:2012 Drinking water specifications to reduce to acceptable limits. Pesticide reduction: Influent challenge of 20–0.01 μg L^{-1} Iron reduction: 3–0.3 mg L^{-1} Fluoride reduction: 6–1 mg L^{-1} NO_3 reduction: 150–45 mg L^{-1}	TDS reduction measured by IS 3025 method (Part 16) Microbial reduction detection by standard membrane filtration (IS10500) method. Metal reduction and pesticide reduction measured using IS 3025 method. SI and pneumatic pressure tests
USEPA guide standard and protocol for testing microbiological water purifiers	Halogenated resins, iodinated resins, ceramic filters, UV-based water treatment systems	Microbial reduction: Bacteria: 6 log at the influent challenge of 10^7 CFU 100 mL^{-1} (*K. terrigena* used) Viruses: 4 log reduction (poliovirus or rotavirus) Protozoa: 2 log reduction (*Giardia lamblia*)	For bacterial testing, spread plate, pour plate, or membrane filtration technique can be used. For viral testing, pour plate or immunofluorescence assay can be used. For protozoan testing, excystation and cyst viability assay used.

(Continued)

Table 8.3 Country-wise protocols for evaluation of PoU water treatment system (*Continued*).

Standard Protocol	Technologies Covered	Reduction Requirements	Methods Described
Canadian protocol for performance testing of drinking water treatment technologies	PoU systems	Microbiological parameters, inorganic substances parameters, organic substances parameters, chlorination by-products, radioactive substance parameters, turbidity parameters should meet the requirements described by local regulations, guidelines for CDWQ, WHO standards, or other generally accepted thresholds.	NSF/ANSI standards
Brazil standard NORM ABNT NBR 15176:2004	Household water purifiers	At least one of the three test criteria must be met: (a) particle retention efficiency, (b) free chlorine removal, or (c) bacterial log10 reduction: *E. coli* log10 decrease of at least 2 logs in water containing 105 CFU mL^{-1}	
Mexico standard PROY-NOM-244-SSA1-2020	Germicidal equipment and substances for domestic water treatment	Mesophilic aerobic bacteria reduction: 95% reduction Total coliform reduction: 4log reduction at the influent challenge of 5000–10 000 CFU mL^{-1}	Chromogenic substrate Coliform test recommended for mesophilic bacteria Standard membrane filtration technique for total coliforms.
China Ministry of Health GB/T 5750-2006 standard and household and similar water treatment units (QB/T 4144-2019) standard, Ministry of Industry and Information Technology		Microbial reduction: (a) 1 log10 reduction at 30–100 CFU mL^{-1} challenge value (b) Reduction of total bacteria plate counts to <100 TPC mL^{-1} at the quarter intervals of the device's lifetime water volume capacity	(a) Plate count method (b) Lactose fermentation test – MPN test.

European standard EN 14897:2006	Disinfection devices, PoU, and POE with low-pressure mercury lamps	Bacterial and phage reductions: Exact log reduction values not specified. Dose–response curve: 10–80 mJ cm^{-2}	NSF/ANSI 55 can be used. For bactericidal activity, data at three points while varying flow rate and UV transmittance has to be collected.
European standard EN 14898:2006 + A1:2007	Both PoU and PoE, plumbed-in only 'active' media filters only	Chlorine reduction: 90% minimum reduction (for class I), or 75%–90% reduction (for class II) at influent 1 mg L^{-1} Organic chemical reduction: As per the requirements described by EUDWD 98/83/EC or WHO guidelines Taste and odour reduction: Based on the reduction of geosmin and 2,4,6-trichlorophenol tested separately (90% reduction from 0.15 to 20 µg L^{-1} challenge, respectively) Lead reduction: 90% reduction at influent challenge of 100 µg L^{-1} Copper reduction: 80% reduction at 3 mg L^{-1} Aluminium reduction: 70% reduction at 600 µg L^{-1} Nitrate reduction: 75% at influent of 200 ± 20 mg L^{-1} NO$_3$	NSF/ANSI 42 or 53 can be used
Europe EN13443-2:2005 + A1:2007	Both PoU and PoE, plumbed-in only mechanical filter	-	NSF/ANSI standard 42 can be used

(Continued)

Table 8.3 Country-wise protocols for evaluation of PoU water treatment system (*Continued*).

Standard Protocol	Technologies Covered	Reduction Requirements	Methods Described
Europe EN 14652:2005 + A1:2007	MF, UF, NF, and RO, both PoE and PoU	Molecular limit for organic compound rejection (UF, NF, RO): At particular concentrations and other test parameters specified by the manufacturer, 90% rejection of the stated capacity of specific organics is necessary. Rate of rejection due to salinity (NF and RO): It is necessary to use 99% of the rated rejection capacity (i.e., tested rejection must be at least 99% of claimed rejection)	NSF/ANSI standard 58
Japanese standard JSA - JIS S 3242	RO	–	–
Japanese standard JSA - JIS K 3835	Bacteriological purifiers	–	–
Japanese standard JSA - JIS S 3201	Household PoU	–	Filtration flow rate test, minimum usable operating water pressure test, recovery test, removal performance test, filtering capability test

Ministry of Health standard has requirements of two challenges for bacteria removal: a 1-log$_{10}$ reduction of *E. coli* resulting in no bacteria in treated test water and a reduction in total plate counts (TPC) of bacteria to less than 100 TPC mL^{-1} in the treated test water. These standards of bacterial (bacteria is considered as microbial target) reduction in treated water are based on health risks, and there is no clear reason assigned for selecting bacteria and achieving the target reductions. However, the People's Republic of China's drinking water quality regulations (GB 5749-2006) include primary standards for *E. coli* and TPCs (Ministry of Health, China).

The standard and protocol solely address bacteria as the only parameter as potential health-related microorganisms of concern and do not contain any viruses or protozoan cysts as target microorganisms. Hence, viruses in drinking water may produce distinct and likely inferior performance effectiveness because such filtration techniques do not target microbial categories other than bacteria. Because protozoa are bigger than the *E. coli* bacteria stated in the protocol, the approach may be adequate in addressing their removal by filtration technology. However, the protocol does not address the potential differential in sensitivity to chemical disinfection processes between *E. coli* bacteria and protozoan parasites. The Chinese Ministry of Industry and Information Technology (MIIT) Standard for PoU and comparable water treatment systems (interestingly, the document number remains the same) recommends total coliform bacteria as the test microorganism and a geometric mean concentration of 200–2000/100 mL in the test water. The complete removal would result in a 2.3-to-3.3-log$_{10}$ reduction. This standard does not indicate any additional test microorganisms. As a result, the bacterial reduction requirements of these various Chinese technological performance evaluation standards and documents are different.

The Ministry of Health standard of China has developed GB/T 5750-2006 to evaluate the different water treatment units (Ministry of Health of the People's Republic of China, 2010). The document has specified two main bacterial challenges. The first criterion is that the PoU water treatment system must exhibit 1 log10 reduction at 30–100 CFU mL^{-1} challenge value. The second criterion is that the treatment unit must reduce total bacteria plate counts to <100 TPC mL^{-1} at the quarter intervals of the device's lifetime water volume capacity. Later in 2020, the MIIT included a geometric mean test for total coliforms in water comprising 200–2000 CFU 100 mL^{-1} or most probable number (MPN)/100 mL. According to the document's microbiological requirements, the log10 reductions would be 2.3–3.3 if the water is expected to be treated to contain 0 total coliforms 100 mL^{-1}.

There are no specific requirements for the virus, a protozoan parasite, or other physicochemical parameter reductions.

8.3.4 European Union

The European Union Standard EN 14652:2005+A1:2007, EN13443-2:2005+A1:2007, EN 14897:2006, EN 14898:2006+A1:2007, etc. specifies requirements relating to the construction, performance, and methods of testing drinking water filters. The standards also include test conditions to evaluate the structural integrity of the units. The standards applicable vary depending on the treatment technologies. Depending on the technology being evaluated, the performance requirements specified are also different for each standard. Generally, chemical reduction and organic molecule rejection capacity are included. EN 14897:2006 also includes bacterial and phage reduction requirements. According to the European standards, the NSF/ANSI protocols can be used for evaluations.

8.3.5 India

BIS is India's national standard body, founded under the BIS Act 2016, for the harmonious development of standardisation, marking, quality certification of goods, and things associated with or incidental thereto. To provide people with safe drinking water, the BIS created drinking water quality standards in India. Drinking water sources must be tested regularly to ensure that the water satisfies the defined requirements and, if not, determine the amount of contamination/unacceptability and the necessary follow-up.

IS 16240: 2015 for RO-based PoU water treatment systems covers RO-based PoU water treatment systems with a capacity of up to 25 L h^{-1} that reduces TDS of water, chemical contamination to a safe level, and removes physical particles such as challenge microorganisms such as bacteria, viruses, and protozoan oocyst Product manual for reverse osmosis (RO) (2015).

Protocols available in various countries are reviewed and presented to highlight significant content, particularly contaminants removal. Most of the protocols specify the removal of challenging microorganisms with little emphasis on chemical contaminants. This also helps provide technology-neutral protocol (applicable to technologies being used in most PoU water treatment systems). In most countries, indicator bacterium such as *E. coli* is used in test water, and the removal efficiency of PoU water treatment systems is determined. However, considering the health risk-based approach, countries are updating the protocols by incorporating the removal of bacteria, viruses, and protozoa through the PoU water treatment systems. Summary of country-wise protocols for evaluating PoU water treatment systems in terms of challenge microorganisms, log 10 removals of these microorganisms etc., are presented in Table 3.2.

8.3.6 Mexico

According to Official Mexico's Standard NOM-244-SSA1-2020 (Official Mexican Standard, 2020) (4-log10 reduction), PoU water treatment technologies/systems must eliminate mesophilic aerobic bacteria by over 95%, and total coliforms reduction should be 99.99%. This standard does not explain the selection of test microorganisms or the reductions in target microorganisms, and the criteria do not appear to be based on health risks (NOM-244-SSA1-2020). Similar to the Brazilian and Chinese standards and protocols, this Mexican protocol solely includes bacteria as potential health-related water-borne pathogens of concern. It concentrates on PoU water filtration systems. As a result, this protocol and standard, like the Brazilian protocol and standard, do not appropriately examine and evaluate filter reduction performance for water-borne enteric viruses. Because protozoans are larger than bacteria like *E. coli*, this methodology and standard may be appropriate for addressing water-borne protozoan removal by filtration processes. However, the Mexican protocol and standard, like the Brazilian protocol and standard, do not address the expected variation in reactions to chemical disinfection methods between *E. coli* bacteria and viruses or protozoa (Bailey *et al.*, 2021).

The Mexican Standard PROY-NOM-244-SSA1-2020 establishes the sanitary requirements and characteristics that equipment and germicidal substances must be met for domestic water treatment. The document includes the protocol for test methods to evaluate efficiency in bacterial reduction. This standard mandates that household treatment methods reduce mesophilic aerobic bacteria by 95% and 4 $\log_{10}$ reductions in total coliform values at 5000–10 000 CFU mL^{-1}. Physicochemical parameters for the influent challenge are not specified, and the viral and oocyst reduction requirements are also not addressed. The target bacterial species have not been provided. However, different methods for detecting mesophilic and coliform bacteria have been described (PROY-NOM-244-SSA1-2016).

8.3.7 The United States of America

The document USEPA guide standard and protocol for testing microbiological water purifiers describes a protocol for evaluating PoU water treatment systems. The United States Environmental Protection Agency (USEPA) is an autonomous executive agency of the United States Federal Government dealing with environmental protection issues. The USEPA (1987) was the first to propose a comprehensive guide standard for PoU water treatment systems that perform microbiological treatment of water. The primary targets of this standard are bacteria, viruses, and disinfectant-resistant parasites (cysts). The USEPA standard protocol utilises realistic worst-case challenges to determine the efficiency of PoU water treatment systems and determine whether the treated water quality meets the requirements as prescribed by US national primary drinking water regulations.

The document covers the protocol for performance evaluation of ceramic filter candles or units, UV disinfection units and halogenated resins. The document outlined microbial reduction performance standards, as well as specific test organisms, microbe concentrations in test waters, and microbial reduction testing procedures. The challenge organisms used were *Klebsiella terrigena* (bacteria), Poliovirus or Rotavirus (virus), and *Giardia lamblia* (cysts). The protocol has also described influent challenge requirements, pH, total organic carbon, TDS and turbidity, based on the evaluated technology. The reduction requirements were 6 log for bacteria, 4 log for viruses, and 2 log for a protozoan. The protocol has also suggested experimental designs and setups, test time conditions and rigs, etc. However, the protocol has stated that the test protocols can be modified depending on the specific testing requirements of the technology.

8.4 CERTIFICATION/EVALUATION PROCESS

Many countries have initiated processes to certify/evaluate PoU water treatment systems. An independent agency normally provides certification through a written assurance that the product meets a specific set of requirements, such as a PoU water treatment system. These requirements are normally well defined and objective parameters against which the system is tested. These requirements are normally termed as standards developed either by an independent agency in individual countries or adapted from international agencies such as the ISO or WHO. However, it is always good to have requirements/standards specific to the countries as these requirements/standards cannot be universally applied. A template for developing the certification/evaluation process is presented below, which can be adopted by any country.

8.4.1 Certification process description

The certification process for drinking water treatment systems/products/units should be open to all manufacturers. The certification should recognise all the water treatment units that adhere to the standards. The rating of the products can also be developed based on the ability and efficiency of the products/units/systems to provide treated water meeting the standards. The rating can be provided depending upon the product performance for decontamination of water for a particular contaminant, the quantity of sludge/reject generated and proposed sludge/reject management.

8.4.1.1 Process development

The process can be developed in accordance with International Organization for Standardization/International Electrotechnical Commission (ISO/IEC) 17067 and guidelines from other leading certification agencies (ISO/IEC 17067:2013). There can be many approaches to certification process development. The following steps can be adopted in certification process development:

- Evaluation of each eligible product based on the review of documents submitted, efficiency testing and quality control protocol adopted by the manufacturer.
- Review of the evaluation outcome by the technical board, which is independent of the evaluation team empowered to grant certification.

The certification process can be technology-specific and coordinated by the Central Agency, preferably a Standard Development Agency in the country. This agency can be a nodal and lead institute for this process. The drinking water treatment products/units/systems can be primarily classified into two groups: household units (PoU and PoE). Further, the technologies can be classified but are not limited to the following categories:

- Adsorption/ion-exchange
- UV radiation
- Membranes (MF, UF, NF, and RO)

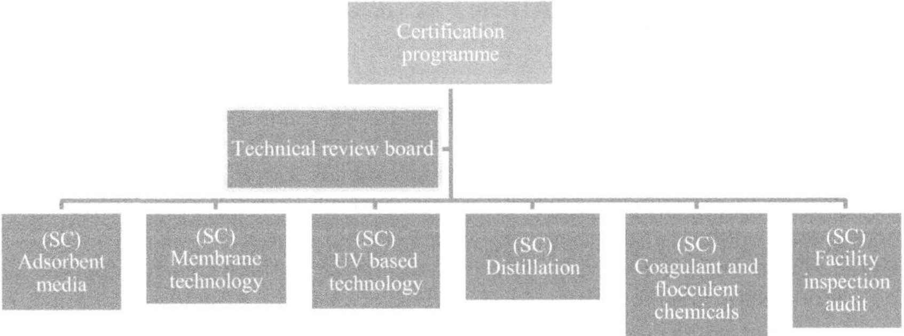

Figure 8.2 Simplified institutional mechanism for certification programme.

- Distillation
- Coagulant and flocculent chemicals

A subcommittee (SC) for each category of technologies can be created, which will act as an advisory body and shall be responsible for the following tasks:

- Finalisation and adaptation of standards
- Defining the evaluation/testing protocol
- Harmonising testing protocol and reporting templates
- Prioritisation criteria for the categories of technologies to be evaluated

The subcommittee should be fully empowered to define, review, and adapt the technical specifications, protocols, and standards to be adopted. A simplified institutional mechanism for the certification programme is presented in Figure 8.2.

8.4.1.2 Technical Review Board

The technical review board (TRB) is the apex authority of the product certification programme. The board can comprise five experts from the field of water technology. Following key responsibilities can be assigned to the TRB:

- Preparation and finalisation of protocol for evaluation of products.
- Review of new applications.
- Decision on grant of certification based on the evaluation outcome.

8.4.1.2.1 Preparation of testing protocol

Protocols are available from International Agencies for testing/evaluation of PoU water treatment systems. Many countries have either initiated developing protocols, or these protocols are already available. However, it is advisable to develop own protocol which can adopt part of already available protocols.

8.4.1.2.2 Review of new applications

TRB can be responsible for reviewing all the new applications. The line of action and type of testing to be conducted to evaluate each product is to be finalised by the TRB. TRB should also decide on the standardised protocol for evaluating the system. The process should be flexible, and the board can invite any subcommittee chair or any other technical expert to finalise the evaluation/ testing protocol if deemed required.

8.4.1.2.3 Preparation and finalisation of standards

TRB can be responsible for preparing the standard for evaluation/testing of each of the technologies mentioned in the earlier section. The standard should include the scope, the detailed procedure to be adopted for evaluation/testing, and the acceptance criteria. These standards may include drinking water quality standards (as these standards are available in most countries, they can be directly adopted, and treated water should meet these standards for specific water quality parameters), test water quality, specific contaminant removal (e.g., bacteria, virus, protozoa) requirements, etc.

8.4.2 Certification process

The certification process should be structured to evaluate each product comprehensively, consistently, and without bias. The process requires gathering technical, administrative, quality control, and sustainability information for the evaluation/testing of each system/technology/product. It is important to note that all the information is critical. The certification process has the following key steps.

8.4.2.1 Application

The application for certification should be filed by the manufacturer (in the case of imported systems/technologies/products, a local agency can apply on the manufacturer's behalf). On scrutiny of the application and relevant documents, if the application is complete and all the required documentation is received, the application shall be forwarded to the technical review team. The manufacturer shall be asked to address the deficiencies in case of any discrepancy before sending the documentation to the review team. The documents submitted with the application must be according to the checklist. When submitting samples to a laboratory for testing, the applicant must always provide parameters. Along with the applications, particular pollutant information, reduction claims, flow rates, capacities, and general product specifications for the products/systems are required. An application fee can be charged along with the application to make the process financially sustainable. The evaluation fee would depend on the extent of work and shall be assessed after the technical review.

The manufacturer needs to submit an entire list of the components of the system/product. Components from secondary or tertiary vendors may be used in certified products. After certification, the manufacturer is responsible for ensuring that all secondary suppliers do not change their material formulations or component sources. Manufacturers who use recognised systems, methods, or goods must have a written agreement with each supplier stating that their component or material composition is used in a product. Changes to a component or formulation must be reported to the certified company and the certification organisation.

8.4.2.2 Document review, preliminary audit, and finalisation of activities

The application number has been assigned once the applicant document is found complete. The application is acknowledged, and a reference number is provided to the applicant for future reference. Following that, the application will be subjected to a technical review. A technical assessment of provided information will ensure the accuracy, completeness, and scope of the certification process. Minimum requirements for certification of the systems/technologies/products making performance claims and the concerned standards would be listed. Depending on the type of systems/technologies/products, the evaluation protocol shall be delineated and followed based on the recommendations of the technical review committee. This may include product claim evaluation, generic treated water quality testing, leaching studies, structural/material integrity, etc.

Manufacturers can acquire certification for materials safety and structural integrity in accordance with the applicable PoU water treatment system regulations. Not all sections from each standard protocol may be required for product certification. This can be the subcommittee's responsibility to ensure and finalise the specific system/product protocol. Technical explanations and suitable system/product bracketing may be used when pursuing certification for several systems/products with comparable functionality. Technical explanations and suitable system/product bracketing may be used. As a result, a technical review will recommend adjustments to see if additional certification-related activities like testing, auditing, or documentation are required. When performance testing is required, specifications will be compared to the original certified material. Standards may be adjusted or waived depending on the system/product type and end-use.

The system/products certified by other certification agencies officially approved as competent to carry out certification by a national accreditation body may have to submit all the documentation related to the evaluation/testing results, certification details, and audit observations. The subcommittee can evaluate test/evaluation data from a laboratory or certification agency on any product being considered for certification. The final protocol and evaluation procedure to be adopted for such products can be entirely at the desecration of the subcommittee. The finalisation of the protocol can be based on the guidelines but does not eliminate judgment based on the experience or technical expertise of the subcommittee.

8.4.2.3 Contract

A contract based on the technical review should be framed and sent to the manufacturer for approval before final activities begin. The contract should list all the activities to be taken as part of the evaluation protocol. It also includes the roles and responsibilities of the manufacturer and the clause of confidentiality related to any proprietary information provided by the manufacturer. The contract should also contain the financial proposal for the product's certification. The contract should include all anticipated certification fees, such as facility assessments, testing, and certification. Before proceeding with certification processes, a deposit may be requested.

This certification programme is only for the final finished systems/products manufactured. The evaluation samples should represent the complete line or set of systems/products to be certified. They should be constructed using similar components and sub-assemblies to those used in production, using production tools and assembled following processes specified for the production run. The certification agency may select the product to be tested from the factory.

8.4.2.4 Evaluation

The system/product is subjected to extensive testing to ensure that it complies with the standard to which it is certified. The testing/evaluation process contains both mandatory and optional testing requirements. Mandatory requirements include the testing of material and structural integrity. For specific technologies, contaminant reduction claims may be mandatory. Moreover, the mandatory testing of the systems and components must be required for certification. The protocol and the extent of evaluation and listing of mandatory and optional testing requirements shall be up to the sub-committee. The exact protocol finalised by the sub-committee shall be used for testing all the systems/products.

Toxicological investigations can be conducted if requested and judged necessary by the subcommittee to determine permissible amounts of pollutants that leach from products during extraction testing. All requirements must be reviewed to assess whether or not a recommendation for certification can be given. When non-conformities have been effectively rectified, the manufacturer must be notified. Non-conformity correspondence may be sent several times during the certification procedure. Product samples are also taken for laboratory testing to determine conformance to standard requirements. The manufacturer must agree to follow a well-defined testing schedule that specifies the precise tests and the frequency with which they should be performed.

A preliminary inspection of the manufacturing facility can be carried out on a mutually agreed date after application. The manufacturing capability and controls, quality control procedures, available testing facilities, and technical skills are all examined during this inspection. Product samples are analysed and drawn for testing at laboratories specific to the system/product. Audits of facilities might also be done every year. The manufacturing of certified systems/products can be assessed to guarantee that the systems on the market are the same tested and certified systems. When deciding if a facility review is required, any factors that doubt the quality system, such as audit history, complaints, or testing history, must be considered. A full facility assessment might involve a review of the manufacturing process and evaluating the quality system. The findings and recommendations can determine how often and for how long surveillance audits should be conducted.

8.4.3 Point-of-use water treatment system testing schedules

The effective water treatment efficiency of PoU water treatment systems varies. Due to functional reductions over time, the performance of the PoU water treatment system may vary with the quantity of water treated for various technologies. The PoU water treatment system testing techniques must be of sufficiently long duration in terms of use time or cumulative quantity of water treated to address changes in flow rate and consequent changes in the system's performance. Flow rate reductions of filters could affect microbial reduction performance, user satisfaction, and willingness to use the filters. In addition to reducing the flow rate of treated water and eventual blockage, some of the filter components may lose their microbial reduction performance efficiency over time and with an increased quantity of water treated. For PoU water treatment systems in which flow may decrease with an increasing quantity of water treated, microbial reduction performance must be identified, quantified, and accounted for, depending on the manufacturer's design criteria and performance claims. As a result, a PoU water treatment system testing plan and schedule for microbiological and water quality composition concerns that appropriately reflect the PoU water treatment system's operating conditions must be addressed.

8.4.3.1 *Testing schedule considerations*

Adequate laboratory testing of a wide range of PoU water treatment systems in a reasonable timeframe, at a reasonable cost, and with representative testing durations demands careful study and preparation. According to some testing techniques and standards the setting should be chosen based on the manufacturer's/vendor's recommendations for long-term operation, usually in terms of the volume of water that can be effectively treated before the system needs to be repaired or components replaced. On the other hand, other protocols define or recommend a more arbitrary test term, usually 14 days, because they have varied performance cycle times, durations, and use circumstances and are designed to have very extended working lifetimes. The entire test time is frequently divided into an appropriate number of equal volume spaced fractions of the PoU water treatment system's total volume life, usually five fractions, to provide test points where the system is challenged with test microorganisms. Some technologies/systems consider low (poor)-quality influent water in some or all volume fractions, which can diminish the microbe reduction effectiveness of the PoU water treatment systems under consideration and constitutes a 'worst-case' scenario. The initial challenge test is performed after a set of replicate PoU water treatment systems has been conditioned or flushed according to the manufacturer's instructions.

Operational water quality measurements in the influent and product water, such as disinfection residuals, turbidity, and pH, are also taken throughout these challenge periods. Several test methods use an easily provided reasonably high-quality water, such as dechlorinated tap water, for up to 16 h day^{-1} to achieve the manufacturer's volume lifetime target in an acceptable amount of time. Some testing methodologies and schedules contain two 48-h stagnation points, the first near the halfway volume mark and the second near the end of the test run total volume, during which no water flows through the PoU water treatment systems. Some forms of PoU water filtration systems are utilised in treatment systems, and a stagnation period may damage the germ removal efficiency. In actual field use of a PoU water treatment system, a prolonged stagnation period may be unrealistic or unreliable, as household members would almost certainly produce treated water almost daily, if not more frequently, negating any beneficial effects from extended water idle times within the filter medium.

8.4.3.2 Communication to the manufacturer

Once the product has undergone performance testing, the technical review committee will review the data, making a certification decision. After evaluation, the recommendation from the subcommittee to certify can be reviewed by the technical review committee, which shall be independent of the evaluation process.

For the certified PoU water treatment system, a certificate of conformity may be prepared and made accessible to the manufacturer. The certification document must include additional claims, verification of performance indicator accuracy, compliance with other standards, and other variations. A licence to use the Standard Mark on a product is only issued once the Technical Review Committee has determined that the manufacturer can consistently manufacture the product according to the relevant standard. The technical review committee can look over all of the findings from the manufacturer's laboratory evaluations, site visits, and quality control protocol to determine the manufacturer's ability to produce goods that meet the relevant standards, particularly in terms of raw materials, manufacturing capability, and quality control facilities such as testing equipment and supervisory staff.

The technical review committee should consider the final grant of certification after scrutiny and verification to its satisfaction.

8.4.3.3 Complaints and appeals

A complaint redressal policy for certification programmes should be designed according to ISO/IEC 17065 to manage product-certification-related complaints from customers, consumers, or other sources. The following are the most common types of complaints:

- Allegations that a corporation that sells certified products has abused the certification mark and/or made false statements about certified products' performance.
- Allegations of a corporation using the certification mark fraudulently to sell products that are not certified.
- Dissatisfaction with the certification procedure or the service quality.

A complaint redressal committee should handle all complaints. When a complaint is received, it should be acknowledged on time. Efforts should be taken to settle the complaint as quickly as possible and keep the complainant informed of the procedure's outcome.

8.4.3.4 Surveillance

Production facility surveillance monitors the quality system's implementation, product literature conformance, and manufacturing repeatability. Each calendar year, companies with certified products can have a minimum of one announced or unannounced surveillance assessment. Based on previous audit history, a comprehensive examination may be done. If any findings do not meet the subcommittee's satisfaction, or if the facility is found to have frequent non-conformities, the TRB may undertake an additional assessment, increase the frequency of audits at the facility, or remove or suspend certification.

Surveillance assessment is an important step in ensuring that the manufacturing facilities adopt a standard protocol or system and that records are traceable. Surveillance assessment of the manufacturing facilities can be carried out independent of evaluation/certification of PoU water treatment systems, which means that a separate team familiar with manufacturing can visit the facilities.

8.5 SUMMARY

Several PoU water treatment systems are developed and introduced in the market with various configurations to address the supply of clean water at the PoU with varying and sometimes unverifiable claims. With the market full of such products and huge claims, it is essential to understand the complexities of water treatment systems and evaluate these claims independently based on the common testing/evaluation/certification protocol. The PoU water treatment products/units/systems evaluation and certification guides consumers, including private and government agencies, in the selection of suitable systems and help the national government directly or indirectly in the number of technical- and evaluation-related functions. The primary objective of certification/evaluation is to provide an independent third-party evaluation of PoU water treatment systems and certify them.

Many international organisations and national governments offer guidance and protocols for evaluating the performance of PoU water treatment systems. The WHO, WQA, and NSF/ANSI are among the most prominent of these, all of which now offer certification or some other type of recommendation based on performance appraisal. In addition, many countries have also established protocols for evaluation/certification of PoU-based water treatment systems for verification of claims. These processes are comprehensive, and certification is granted after rigorous laboratory-based evaluation, and at time surveillance of the manufacturing facilities is also undertaken.

Most of these protocols are meant to verify claims for microorganisms' removal and require log removal of bacteria, viruses and protozoa are included in these protocols. These protocols, particularly from international agencies, are fairly comprehensive in defining test water composition, the number of water samples, required log removal of representative microorganisms, number of test runs, etc., including traceability of the tests. Considering the importance of certification, which also help manufacturers in properly marketing their PoU water systems, the importance of getting their systems certified is growing them.

In addition, several national governments are also coming forward to develop their protocols for verification of claims of the PoU water treatment systems being sold in the country. A template for developing a certification/evaluation process is presented in this chapter which any country can adopt.

References

Ang W. L. and Mohammad A. W. (2015). Mathematical modeling of membrane operations for water treatment. In: A. Basile, A. Cassano, and N. K. Rastogi (eds.), Advances in Membrane Technologies for Water Treatment, Woodhead Publishing, United Kingdom, pp. 379–407, https://doi.org/10.1016/C2013-0-16469-0

Arnal J. M., Garcia-Fayos B. and Sancho M. (2011). Membrane cleaning. In: R. Y. Ning (ed.), Expanding Issues in Desalination, IntechOpen, pp. 63–84, https://doi.org/10.5772/19760

Bailey E. S., Beetsch N., Wait D. A., Oza H. H., Ronnie N. and Sobsey M. D. (2021). Methods, protocols, guidance and standards for performance evaluation for point-of-use water treatment technologies: history, current status, future needs and directions. *Water*, **13**(8), 1094, https://doi.org/10.3390/w13081094

Bain R., Cronk R., Wright J., Yang H., Slaymaker T. and Bartram J. (2014). Fecal contamination of drinking-water in low- and middle-income countries: a systematic review and meta-analysis. *PLoS Medicine*, **11**(5), e1001644, https://doi.org/10.1371/journal.pmed.1001644

Basile A., Cassano A. and Rastogi N. K. (eds.). (2015). Advances in Membrane Technologies for Water Treatment: Materials, Processes and Applications. Elsevier, United Kingdom.

Beca Limited (2010). CH2M Beca Limited (2010) Drinking Water Standards New Zealand Cost Benefit Analysis-Engineering Input.

Brandt M. J., Johnson K. M., Elphinston A. J. and Ratnayaka D. D. (2016). Twort's Water Supply. Butterworth-Heinemann, Oxford, UK.

Canu I. G., Laurent O., Pires N., Laurier D. and Dublineau I. (2011). Health effects of naturally radioactive water ingestion: the need for enhanced studies. *Environmental Health Perspectives*, **119**(12), 1676–1680, https://doi.org/10.1289/ehp.1003224

Chatterley C. and Linden K. (2010). Demonstration and evaluation of germicidal UV-LEDs for point-of-use water disinfection. *Journal of Water and Health*, **8**(3), 479–486, https://doi.org/10.2166/wh.2010.124

Chaturvedi S. and Dave P. N. (2012). Removal of iron for safe drinking water. *Desalination*, **303**, 1–11, https://doi.org/10.1016/j.desal.2012.07.003

CH2M Beca Limited (2010). Drinking Water Standards New Zealand Cost Benefit Analysis-Engineering Input.

Cunha M. D. C. and Sousa J. (1999). Water distribution network design optimization: simulated annealing approach. *Journal of Water Resources Planning and Management*, **125**(4), 215–221, https://doi.org/10.1061/(ASCE)0733-9496(1999)125:4(215)

Elala D., Labhasetwar P. and Tyrrel S. F. (2011). Deterioration in water quality from supply chain to household and appropriate storage in the context of intermittent water supplies. *Water Science and Technology: Water Supply*, **11**(4), 400–408, https://doi.org/10.2166/ws.2011.064

Ercumen A., Naser A. M., Unicomb L., Arnold B. F., Colford Jr. J. M. and Luby S. P. (2015). Effects of source-versus household contamination of tubewell water on child diarrhea in rural Bangladesh: a randomised controlled trial. *PloS One*, **10**(3), e0121907, https://doi.org/10.1371/journal.pone.0121907

Fane A. G., Wang R. and Jia Y. (2011). Membrane technology: past, present and future. In: Membrane and Desalination Technologies, Humana Press, Totowa, NJ, pp. 1–45.

Field R. (2010). Fundamentals of fouling. *Membranes for Water Treatment*, **4**, 1–23.

Haidari A. H., Heijman S. G. J. and Van Der Meer W. G. J. (2018). Optimal design of spacers in reverse osmosis. *Separation and Purification Technology*, **192**, 441–456, https://doi.org/10.1016/j.seppur.2017.10.042

Hamouda M. A., Anderson W. B. and Huck P. M. (2010). A framework for selecting POU/POE systems. *Journal-American Water Works Association*, **102**(12), 42–56, https://doi.org/10.1002/j.1551-8833.2010.tb11362.x

Harris J. (2005). Challenges to the Commercial Viability of Point-of-Use (POU) Water Treatment Systems in Low-Income Settings. Master Thesis. Oxford School of Geography and the Environment Oxford University.

Hassan A. R., Abdull N. and Ismail A. F. (2007). A theoretical approach on membrane characterization: the deduction of fine structural details of asymmetric nanofiltration membranes. *Desalination*, **206**(1–3), 107–126, https://doi.org/10.1016/j.desal.2006.06.008

Holmes C., Cervia J. S., Ortolano G. A. and Canonica F. P. (2010). Preventive efficacy and cost-effectiveness of point-of-use water filtration in a subacute care unit. *American Journal of Infection Control*, **38**(1), 69–71, https://doi.org/10.1016/j.ajic.2009.04.284

Ismail F., Khulbe K. C. and Matsuura T. (2018). Reverse Osmosis. Elsevier, Geneva, Switzerland, https://doi.org/10.1016/C2016-0-00580-6

ISO/IEC 17067:2013. Fundamentals of product certification and guidelines for product certification schemes, International Organization for Standardization/International Electrotechnical Commission.

John V., Jain P., Rahate M. and Labhasetwar P. (2014). Assessment of deterioration in water quality from source to household storage in semi-urban settings of developing countries. *Environmental Monitoring and Assessment*, **186**(2), 725–734, https://doi.org/10.1007/s10661-013-3412-z

Jolley R. L., Brungs W. A., Cotruvo J. A., Cumming R. B., Mattice J. S. and Jacobs V. A. (1983). Water chlorination: environmental impact and health effects. **4**(2), Environment, health, and risk. United States.

Kavianipour O., Ingram G. D. and Vuthaluru H. B. (2019). Studies into the mass transfer and energy consumption of commercial feed spacers for RO membrane modules using CFD: effectiveness of performance measures. *Chemical Engineering Research and Design*, **141**, 328–338, https://doi.org/10.1016/j.cherd.2018.10.041

Kedem O. and Katchalsky A. (1958). Thermodynamic analysis of the permeability of biological membranes to non-electrolytes. *Biochimica et Biophysica Acta*, **27**, 229–246, https://doi.org/10.1016/0006-3002(58)90330-5

Krishnan J. N., Venkatachalam K. R., Ghosh O., Jhaveri K., Palakodeti A. and Nair N. (2022). Review of thin film nanocomposite membranes and their applications in desalination. *Frontiers in Chemistry*, **10**, Article ID: 781372, https://doi.org/10.3389/fchem.2022.781372

Kuiper S., Van Rijn C. J. M., Nijdam W. and Elwenspoek M. C. (1998). Development and applications of very high flux microfiltration membranes. *Journal of Membrane Science*, **150**(1), 1–8, https://doi.org/10.1016/S0376-7388(98)00197-5

Lytle D. A., Williams D., Muhlen C., Riddick E. and Pham M. (2020). The removal of ammonia, arsenic, iron and manganese by biological treatment from a small Iowa drinking water system. *Environmental Science: Water Research and Technology*, **6**(11), 3142–3156, https://doi.org/10.1039/D0EW00361A

Ma S., Song L., Ong S. L., & Ng W. J. (2004). A 2-D streamline upwind Petrov/Galerkin finite element model for concentration polarization in spiral wound reverse osmosis modules. *Journal of Membrane Science*, **244**(1–2), 129–139.

Maddah H. and Chogle A. (2017). Biofouling in reverse osmosis: phenomena, monitoring, controlling and remediation. *Applied Water Science*, **7**(6), 2637–2651, https://doi.org/10.1007/s13201-016-0493-1

Malkoske T. A., Bérubé P. R. and Andrews R. C. (2020). Coagulation/flocculation prior to low pressure membranes in drinking water treatment: a review. *Environmental Science: Water Research and Technology*, **6**(11), 2993–3023, https://doi.org/10.1039/D0EW00461H

McMordie Stoughton K., Duan X. and Wendel E. M. (2013). Reverse Osmosis Optimization (No. PNNL-22682). Pacific Northwest National Lab. (PNNL), Richland, WA, United States.

Medeiros R. C., de MN Fava N., Freitas B. L. S., Sabogal-Paz L. P., Hoffmann M. T., Davis J. and … Byrne J. A. (2020). Drinking water treatment by multistage filtration on a household scale: efficiency and challenges. *Water Research*, **178**, 115816, https://doi.org/10.1016/j.watres.2020.115816

Medema G. J., Shaw S., Waite M., Snozzi M., Morreau A. and Grabow W. (2003). Catchment characterisation and source water quality. *Assessing Microbial Safety of Drinking Water*, **4**, 111–158.

Ministry of Health of the People's Republic of China (2010). Microbial Standard for Drinking Water Purifier; GB/T 5750-2006. Ministry of Health of the People's Republic of China, Beijing, China.

Mintz E., Bartram J., Lochery P. and Wegelin M. (2001). Not just a drop in the bucket: expanding access to point-of-use water treatment systems. *American Journal of Public Health*, **91**(10), 1565–1570, https://doi.org/10.2105/AJPH.91.10.1565

Mohsin M., Safdar S., Asghar F. and Jamal F. (2013). Assessment of drinking water quality and its impact on residents' health in Bahawalpur city. *International Journal of Humanities and Social Science*, **3**(15), 114–128.

Moore D., Poynton M., Broadbent J. M. and Thomson W. M. (2017). The costs and benefits of water fluoridation in NZ. *BMC Oral Health*, **17**(1), 1–8, https://doi.org/10.1186/s12903-017-0433-y

Nair R. R., Protasova E., Strand S. and Bilstad T. (2018). Implementation of Spiegler–Kedem and steric hindrance pore models for analyzing nanofiltration membrane performance for smart water production. *Membranes*, **8**(3), 78, https://doi.org/10.3390/membranes8030078

Nelligan T., Wirth S., De Cuypere C., Mach L. (2015). Operation and maintenance costs of drinking water plants. https://www150.statcan.gc.ca/n1/pub/16-002-x/2011001/part-partie3-eng.htm

Nguyen T., Roddick F. A. and Fan L. (2012). Biofouling of water treatment membranes: a review of the underlying causes, monitoring techniques and control measures. *Membranes*, **2**(4), 804–840, https://doi.org/10.3390/membranes2040804

Ning R. Y. (2012). Chemistry in the operation and maintenance of reverse osmosis systems. *Chap*, **4**, 85–95.

Official Mexican Standard (2020). NOM-244-SSA1 Equipment and germicidal substances for domestic water treatment. Sanitary requirements, Mexico City, Mexico.

NSF International (2012). NSF Protocol P248.01: Military Operations Microbiological Water Purifiers. NSF International, Ann Arbor, MI, USA.

NSF International. (2014). NSF Protocol P231: Microbiological Water Purifiers. NSF International, Ann Arbor, MI, USA.

NSF International. (2019a). NSF Protocol 62-2019: Drinking Water Distillation Systems. NSF International, Ann Arbor, MI, USA.

NSF International. (2019b). NSF Protocol 55-2019: Ultraviolet Microbiological Water Treatment Systems. NSF International, Ann Arbor, MI, USA.

NSF International. (2019c). NSF Standard 244-2019: Supplemental Municipal Water Treatment Systems – Filtration. NSF International, Ann Arbor, MI, USA.

NSF International. (2019d). NSF Protocol 53-2019: Drinking Water Treatment Units – Health Effects. NSF International, Ann Arbor, MI, USA.

NSF International. (2019e). NSF Protocol 58-2019: Reverse Osmosis Drinking Water Treatment Systems. NSF International, Ann Arbor, MI, USA.

Okpara C. G., Oparaku N. F. and Ibeto C. N. (2011). An overview of water disinfection in developing countries and potentials of renewable energy. *Journal of Environmental Science and Technology*, **4**(1), 18–30, https://doi.org/10.3923/jest.2011.18.30

Omer N. H. (2019). Water quality parameters. Water quality – science, assessments and policy, 18.

Pagsuyoin S. A., Santos J. R., Latayan J. S., and Barajas J. R. (2015). A multi-attribute decision-making approach to the selection of point-of-use water treatment. *Environment Systems and Decisions*, **35**(4), 437–452, https://doi.org/10.1007/s10669-015-9567-0

Palansooriya K. N., Yang Y., Tsang Y. F., Sarkar B., Hou D., Cao X., Meersg E., Rinklebeh J., Kim K.H., Ok Y. S. (2020). Occurrence of contaminants in drinking water sources and the potential of biochar for water quality improvement: a review. *Critical Reviews in Environmental Science and Technology*, **50**(6), 549–611, https://doi.org/10.1080/10643389.2019.1629803

Pandey S. R., Jegatheesan V., Baskaran K. and Shu L. (2012). Fouling in reverse osmosis (RO) membrane in water recovery from secondary effluent: a review. *Reviews in Environmental Science and Bio/Technology*, **11**(2), 125–145, https://doi.org/10.1007/s11157-012-9272-0

Peppas N. A. and Meadows, D. L. (1983). Macromolecular structure and solute diffusion in membranes: an overview of recent theories. *Journal of Membrane Science*, **16**, 361–377.

Pappenheimer J. R. (1953). Passage of molecules through capillary walls. *Physiological Reviews*, **33**(3), 387–423, https://doi.org/10.1152/physrev.1953.33.3.387

Patel A. I., Hecht C. E., Cradock A., Edwards M. A. and Ritchie L. D. (2020). Drinking water in the United States: implications of water safety, access, and consumption. *Annual Review of Nutrition*, **40**, 345–373, https://doi.org/10.1146/annurev-nutr-122319-035707

Peter-Varbanets M., Zurbrügg C., Swartz C. and Pronk W. (2009). Decentralized systems for potable water and the potential of membrane technology. *Water Research*, **43**(2), 245–265, https://doi.org/10.1016/j.watres.2008.10.030

Product manual for reverse osmosis (RO). (2015). Product manual for reverse osmosis (RO) based on point-of-use (PoU) water treatment system according to IS 16240.

Ren D., Colosi L. M. and Smith J. A. (2013). Evaluating the sustainability of ceramic filters for point-of-use drinking water treatment. *Environmental Science and Technology*, **47**(19), 11206–11213, https://doi.org/10.1021/es4026084

Renkin E. M. (1954). Filtration, diffusion, and molecular sieving through porous cellulose membranes. *The Journal of General Physiology*, **38**(2), 225.

Rivera-Utrilla J., Sánchez-Polo M., Gómez-Serrano V., Álvarez P. M., Alvim-Ferraz M. C. M. and Dias J. M. (2011). Activated carbon modifications to enhance its water treatment applications. An overview. *Journal of Hazardous Materials*, **187**(1–3), 1–23, https://doi.org/10.1016/j.jhazmat.2011.01.033

Robertson B. C. and Zydney, A. L. (1988). A Stefan-Maxwell analysis of protein transport in porous membranes. *Separation Science and Technology*, **23**(12–13), 1799–1811.

Rogers E., Tappis H., Doocy S., Martínez K., Villeminot N., Suk A., Kumar D., Pietzsch S., Puett C. (2019). Costs and cost-effectiveness of three point-of-use water treatment technologies added to community-based treatment of severe acute malnutrition in Sindh Province, Pakistan. *Global Health Action*, **12**(1), 1568827, https://doi.org/1 0.1080/16549716.2019.1568827

Roy A., Moulik S., Kamesh R. and Mullick A. (eds). (2020). Modeling in Membranes and Membrane-Based Processes. John Wiley and Sons, Beverly, Massachusetts.

Sala-Garrido R., Mocholi-Arce M., Molinos-Senante M. and Maziotis A. (2021). Marginal abatement cost of carbon dioxide emissions in the provision of urban drinking water. *Sustainable Production and Consumption*, **25**, 439–449, https:// doi.org/10.1016/j.spc.2020.11.025

Sander A., Berghult B., Broo A. E., Johansson E. L. and Hedberg T. A. (1996). Iron corrosion in drinking water distribution systems – the effect of pH, calcium and hydrogen carbonate. *Corrosion Science*, **38**(3), 443–455, https://doi. org/10.1016/0010-938X(96)00142-4

Schwinge J., Neal P. R., Wiley D. E., Fletcher D. F. and Fane A. G. (2004). Spiral wound modules and spacers: review and analysis. *Journal of Membrane Science*, **242**(1–2), 129–153, https://doi.org/10.1016/j.memsci.2003.09.031

Seifert-Dähnn I., Nesheim I., Gosh S., Dhawde R., Ghadge A. and Wennberg A. C. (2017). Variations of drinking water quality influenced by seasons and household interventions: a case study from rural Maharashtra, India. *Environments*, **4**(3), 59, https://doi.org/10.3390/environments4030059

Shields K. F., Bain R. E., Cronk R., Wright J. A. and Bartram J. (2015). Association of supply type with fecal contamination of source water and household stored drinking water in developing countries: a bivariate meta-analysis. *Environmental Health Perspectives*, **123**(12), 1222–1231, https://doi.org/10.1289/ehp.1409002

Sillanpää M., Ncibi M. C., Matilainen A. and Vepsäläinen M. (2018). Removal of natural organic matter in drinking water treatment by coagulation: A comprehensive review. *Chemosphere*, **190**, 54–71, https://doi.org/10.1016/j.chemosphere.2017.09.113

Singh C. P., Yadav A., and Kumar A. (2022). Numerical simulations of the effect of spacer filament geometry and orientation on the performance of the reverse osmosis process. *Colloids and Surfaces A: Physicochemical and Engineering Aspects*, **650**, 129664, https://doi.org/10.1016/j.colsurfa.2022.129664

Strathmann H., Giorno L. and Drioli E. (2011). Introduction to Membrane Science and Technology. Wiley-VCH, Weinheim, Germany, Vol. **544**.

USEPA. (1987). Guide Standard and Protocol for Testing Microbiological Water Purifiers.

Verhougstraete M., Reynolds K. A., Pearce-Walker J. and Gerba C. (2020). Cost-benefit analysis of point-of-use devices for health risks reduction from pathogens in drinking water. *Journal of Water and Health*, **18**(6), 968–982, https://doi. org/10.2166/wh.2020.111

Verniory A., Du Bois R., Decoodt P., Gassee J. P. and Lambert P. P. (1973). Measurement of the permeability of biological membranes application to the glomerular wall. *The Journal of General Physiology*, **62**(4), 489–507, https://doi.org/10.1085/jgp.62.4.489

Wang X. C. and Luo L. (2021). Water cycle management for building water-wise cities" in Water-Wise Cities and Sustainable Water Systems: Concepts, Technologies, and Applications. IWA Publishing, 151–179, https://doi.org/10.2166/9781789060768_0151

Water Quality Association Annual Education Kit. Water Quality Association Annual Education Kit Volume X – Article 4.

World Health Organization (2020). WHO global water, sanitation and hygiene: annual report 2019.

World Health Organization (2022). Guidelines for drinking-water quality-Fourth edition incorporating the first and second addenda.

World Health Organization and World Health Organisation Staff (2004). Guidelines for Drinking-Water Quality. World Health Organization, Geneva, Switzerland, Vol. **1**.

World Health Organization (2011). Evaluating household water treatment options: Health-based targets and microbiological performance specifications. World Health Organization.

WQRF. (2021). WQRF (2021) Water quality Research Fund Annual report 2020.

Wright J., Gundry S. and Conroy R. (2004). Household drinking water in developing countries: a systematic review of microbiological contamination between source and point-of-use. *Tropical Medicine and International Health*, **9**(1), 106–117, https://doi.org/10.1046/j.1365-3156.2003.01160.x

Wu J., Cao M., Tong D., Finkelstein Z. and Hoek E. (2021). A critical review of point-of-use drinking water treatment in the United States. *NPJ Clean Water*, **4**(1), 1–25, https://doi.org/10.1038/s41545-020-00095-x

Yadav P. R. (2006). Environmental Biotechnology. Discovery Publishing House, New Delhi, India.

Yin R. and Shang C. (2020). Removal of micropollutants in drinking water using UV-LED/ chlorine advanced oxidation process followed by activated carbon adsorption. *Water Research*, **185**, 116297, https://doi.org/10.1016/j.watres.2020.116297

Yu T., Meng L., Zhao Q. B., Shi Y., Hu H. Y. and Lu Y. (2017). Effects of chemical cleaning on RO membrane inorganic, organic and microbial foulant removal in a full-scale plant for municipal wastewater reclamation. *Water Research*, **113**, 1–10, https://doi.org/10.1016/j.watres.2017.01.068

Zhang K., Achari G., Sadiq R., Langford C. H. and Dore M. H. (2012). An integrated performance assessment framework for water treatment plants. *Water Research*, **46**(6), 1673–1683, https://doi.org/10.1016/j.watres.2011.12.006

Website

www.cyber-nook.com/chart/default.asp.

www.fixr.com/costs/water-purification-system

www.lenntech.com/membrane-cleaning.htm

www.marketsandmarkets.com/Market-Reports/point-of-use-water-treatment-systems-market-131277828.html

www.mordorintelligence.com/industry-reports/membrane-water-and-wastewater-treatment-wwt-market

www.purewaterproducts.com

www.un.org/sustainabledevelopment/water-and-sanitation

Index

IWA Publishing's authorised EU representative for General Product Safety Regulations is Diane D'Arras, 15 rue Duret, 75116 Paris, France, e-mail: safety@iwap.co.uk.

Printed and bound by CPI Group (UK) Ltd, Croydon, CR0 4YY
16/06/2026
02140369-0001